Good Reception

The John D. and Catherine T. MacArthur Foundation Series on Digital Media and Learning

Civic Life Online: Learning How Digital Media Can Engage Youth, edited by W. Lance Bennett

Digital Media, Youth, and Credibility, edited by Miriam J. Metzger and Andrew J. Flanagin

Digital Youth, Innovation, and the Unexpected, edited by Tara McPherson

The Ecology of Games: Connecting Youth, Games, and Learning, edited by Katie Salen

Learning Race and Ethnicity: Youth and Digital Media, edited by Anna Everett

Youth, Identity, and Digital Media, edited by David Buckingham

Engineering Play: A Cultural History of Children's Software, by Mizuko Ito

Hanging Out, Messing Around, and Geeking Out: Kids Living and Learning with New Media, by Mizuko Ito et al.

The Civic Web: Young People, the Internet, and Civic Participation, by Shakuntala Banaji and David Buckingham

Connected Play: Tweens in a Virtual World, by Yasmin B. Kafai and Deborah A. Fields

The Digital Youth Network: Cultivating Digital Media Citizenship in Urban Communities, edited by Brigid Barron, Kimberley Gomez, Nichole Pinkard, and Caitlin K. Martin

The Interconnections Collection, developed by Kylie Peppler, Melissa Gresalfi, Katie Salen Tekinbaş, and Rafi Santo

- *Gaming the System: Designing with Gamestar Mechanic*, by Katie Salen Tekinbaş, Melissa Gresalfi, Kylie Peppler, and Rafi Santo
- *Script Changers: Digital Storytelling with Scratch*, by Kylie Peppler, Rafi Santo, Melissa Gresalfi, and Katie Salen Tekinbaş
- *Short Circuits: Crafting E-Puppets with DIY Electronics*, by Kylie Peppler, Katie Salen Tekinbaş, Melissa Gresalfi, and Rafi Santo
- *Soft Circuits: Crafting E-Fashion with DIY Electronics*, by Kylie Peppler, Melissa Gresalfi, Katie Salen Tekinbaş, and Rafi Santo

Connected Code: Children as the Programmers, Designers, and Makers for the Twenty-first Century, by Yasmin B. Kafai and Quinn Burke

Disconnected: Youth, New Media, and the Ethics Gap, by Carrie James

Education and Social Media: Toward a Digital Future, edited by Christine Greenhow, Julia Sonnevend, and Colin Agur

Framing Internet Safety: The Governance of Youth Online, by Nathan W. Fisk

Connected Gaming: What Making Video Games Can Teach Us about Learning and Literacy, by Yasmin B. Kafai and Quinn Burke

Giving Voice: Mobile Communication, Disability, and Inequality, by Meryl Alper

Digital Expectations: A Generation at Risk and Why We're Worried about the Wrong Things, by Jacqueline Ryan Vickery

Good Reception: Teens, Teachers, and Mobile Media in a Los Angeles High School, by Antero Garcia

Good Reception

Teens, Teachers, and Mobile Media in a Los Angeles High School

Antero Garcia

The MIT Press
Cambridge, Massachusetts
London, England

This book was set in ITC Stone Sans Std and ITC Stone Serif Std by Toppan Best-set Premedia Limited.

Library of Congress Cataloging-in-Publication Data

Names: Garcia, Antero, author.
Title: Good reception : teens, teachers, and mobile media in a Los Angeles high school / Antero Garcia.
Description: Cambridge, MA : MIT Press, [2017] | Series: The John D. and Catherine T. MacArthur Foundation series on digital media and learning | Includes bibliographical references and index.
Identifiers: LCCN 2017011034 | ISBN 9780262037082 (hardcover : alk. paper) ISBN 9780262545785 (paperback)
Subjects: LCSH: Mobile communication systems in education--California--Los Angeles. | Media literacy--Study and teaching (Secondary)--California--Los Angeles. | Education, Secondary--California--Los Angeles.
Classification: LCC LB1044.84 .G36 2017 | DDC 371.33--dc23 LC record available at https://lccn.loc.gov/2017011034

This book is dedicated to the memory of my father, Jose "Joey" Angel Garcia. He taught me to play, to question, to dream, and to laugh. Thank you. I miss you.

Contents

Series Foreword

In recent years, digital media and networks have become embedded in our everyday lives and are part of broad-based changes in how we engage in knowledge production, communication, and creative expression. Unlike the early years in the development of computers and computer-based media, digital media are now commonplace and pervasive, having been taken up by a wide range of individuals and institutions in all walks of life. Digital media have escaped the boundaries of professional and formal practice and the academic, governmental, and industry homes that initially fostered their development. Now, diverse populations and non-institutionalized practices, including the peer activities of youths, have embraced them. Although specific forms of technology uptake are highly diverse, a generation is growing up in an era when digital media are part of the taken-for-granted social and cultural fabric of learning, play, and social communication.

This book series is founded on the working hypothesis that those immersed in new digital tools and networks are engaged in an unprecedented exploration of language, games, social interaction, problem solving, and self-directed activity that leads to diverse forms of learning. These diverse forms of learning are reflected in expressions of identity, in the ways that individuals express independence and creativity, and in those individuals' ability to learn, exercise judgment, and think systematically.

The defining frame for this series is not a particular theoretical or disciplinary approach, nor is it a fixed set of topics. Rather, the series revolves around a constellation of topics investigated from multiple disciplinary and practical frames. The series as a whole looks at the relationship between youth, learning, and digital media, but each contribution might deal with only a subset of this constellation. Erecting strict topical boundaries would exclude some of the most important work in the field. For example, restricting the content of the series only to people of a certain age would mean artificially reifying an age boundary when the phenomenon demands otherwise. This would become especially problematic with new forms of online participation where one crucial outcome is the mixing of participants of different ages. The same goes for digital media, which are increasingly inseparable from analog and earlier media forms.

The series responds to certain changes in our media ecology that have important implications for learning. These changes involve new forms of media literacy and developments in the modes of media participation. Digital media are part of a convergence between interactive media (most notably gaming), online networks, and existing media forms. Navigating this media ecology involves a palette of literacies that are being defined through practice yet require more scholarly scrutiny before they can be fully incorporated pervasively into educational initiatives. Media literacy involves not only ways of understanding, interpreting, and critiquing media but also the means for creative and social expression, online search and navigation, and a host of new technical skills. The potential gap in literacies and participation skills creates new challenges for educators who struggle to bridge media engagement inside and outside the classroom.

The John D. and Catherine T. MacArthur Foundation Series on Digital Media and Learning, published by the MIT Press, aims to close these gaps and provide innovative ways of thinking about and using new forms of knowledge production, communication, and creative expression.

Acknowledgments

This work would not exist without the voices, dreams, and intellectual contributions of the students of the school I refer to as South Central High School. Although my name appears on the front of this book, my students' insights and knowledge fundamentally drive the work found in these pages, and I thank you all for challenging me and helping me grow as a researcher, teacher, and learner. I am honored to have worked with all of you. I also want to thank the teachers of South Central High School for the work you have done for the families of South Central Los Angeles. I continue to be amazed by the energy you invest in our school and in the lives of our students. In particular, I want to thank the many collaborators who helped create the Schools for Community Action: your vision of a renewed and revitalized learning experience for South Central fills me with optimism. I am honored to have worked with each of you. Much of the work in this book was originally conceived as the culmination of doctoral work while studying at the University of California, Los Angeles. I want to acknowledge the guidance and mentorship of my dissertation committee—Ernest Morrell, Marjorie Faulstich Orellana, Greg Niemeyer, and John Rogers. I am grateful for all of your time and support. Likewise, several other professors—including Kris Gutiérrez, Thomas Philip, Rema Reynolds, and Alfred Tatum—gave me crucial feedback on this work. Thank you all for your support. Clifford Lee, Danny Martinez, and Nicole Mira have heard me talk about the work featured in this book ad nauseam and through conversations and feedback helped me refine my thinking. Thank you for helping to advance a shared vision of equity through your work, as well. I also want to thank Susan Buckley for patiently ushering this work closer and closer to presentable shape. I don't even fault you for being a member of the Enlightened Ingress faction. I also want to recognize the amazing educators and organizers in Los Angeles who helped spearhead the Schools for Community Action and who have helped me stay honest in my commitment to equity-driven change in urban schools. Peter Carlson, Samantha Diego, Mark Gomez, Patricia Hanson, Travis Miller, Katie Rainge-Briggs, Tony Terry, and the many other teachers, parents, students, and South Central community members: thank you for your daily commitment to the students in South Central Los Angeles.

Considering the amount of time that I have spent contemplating, looking at, and writing about mobile devices, one would think that I would be better at calling and responding to text messages. In any case, Mom, Angela, and Kyoko: thanks for your support. Finally, Ally, thank you for your patience and love. This work would not exist without you.

Introduction

Missing Class

It was a Wednesday in late May 2011, twenty minutes after noon. My stomach growled, a reminder that—as usual—I'd skipped breakfast in my haste to battle traffic and get to school early enough to wait in line at the usually broken copy machine and to get things ready for my ninth graders. In all likelihood, some of my students' stomachs were growling as well. Many students came to school without eating breakfast.

But my students weren't in the room with me at the moment. In fact, I had lost my class. I don't mean that in the figurative way, where teachers lose control of a class and silently pray for the bell to ring even a few moments earlier. I mean *literally* I had lost my class: I had no idea where the students in my third period currently were. Any of them.

I checked my phone anxiously, wondering if one or two of them would text me to check in. I also checked the voicemail for Anansi, the mythological spider god who had been participating in our class. He'd been around quite a bit lately, and I wondered if a student had made a last-ditch call asking the trickster for some help.

Just as I was giving up hope and hanging my head in shame (a veteran teacher who couldn't even keep track of his students), Elizabeth flung open the classroom door. She was slightly out of breath, and her eyes were wide in excitement.

"Yo, Garcia. I found some! Dante got to a few before me, but I had my own plan and started on the far end of the school." She waved a handful of crumpled papers, her iPod, and some dangling plastic spiders in my general direction.

While I was worrying that some of my students might find themselves in trouble for walking around the school without the proper documentation, my students were searching, recording, and analyzing social and physical components of our school space. They didn't know it yet, but by the time they all were back in class (and they all did come back; Elizabeth renewed my faith in them), these students would be ready to begin transforming their school community. They would use not complex materials but paper, tape, iPods, and plastic spiders. And by the time they were done, Elizabeth and her classmates had reinterpreted the school and publicly documented some powerful counternarratives to guide social change at South Central High School.

This is a book about kids in schools in the twenty-first century. Their experiences of being in schools, learning, and engaging with technology are fundamentally different than my own when I was their age. I grew up a child of the 1980s. As I was sitting in classrooms, I didn't know that I was witnessing the beginning of three decades of rigorous educational reform. I was an infant the year that *A Nation at Risk: The Imperative for Educational Reform* was released in 1983 and oblivious to the changes that the report ushered in. Ringing a warning cry or death knell for public education in the United States, the report famously noted:

> If an unfriendly foreign power had attempted to impose on America the mediocre educational performance that exists today, we might well have viewed it as an act of war. As it stands, we have allowed this to happen to ourselves.[1]

The report triggered many of the sweeping reforms that exist in the educational landscape today. And yet, for all of the hand wringing about how to fix schools and education, few of these reforms have actually addressed how significantly the world has changed. Despite the fact that kids today are regularly engaged in learning—outside of schools, on digital devices, and with friends—in ways that simply were not possible a decade or two ago, schools still largely operate in the same ways. Sure, the tools inside of schools are getting shinier, digital, and more expensive to keep up to date, but schools still plod along in the same traditional format.

In this book, I want to tell you the story of my students, their experiences with an increasingly out-of-date school setting, and the ways that trust and school relationships can push change in classrooms. Over the course of a year, I explored how the young people in one American school are learning and socializing differently than they did in even the recent past, what these changes mean for school policies, and what teachers can do to meet kids' needs today. A teacher, I put my classroom and my school at the center of this discussion. I am not a perfect teacher, and South Central High School (SCHS) is not a perfect school and has been labeled a "dropout factory."[2] However, if we are going to look at how students *actually* engage in schools today, we need to look closely at an actual school and actual students, warts and all. For most of this book, I draw on incidents and challenges I faced while working at SCHS. My experiences as a language arts teacher in one of the oldest and most structurally inconsistent schools in southern California provide a perspective on technology, literacy, and education in urban schools that often is lacking in public discourse about public education in the twenty-first century.

We are due to make radical changes to schools. Teachers who are working closely with an "always-connected generation"[3] have an opportunity to guide how technology and cultural shifts are taken up by students. For instance, I wrote most of this introduction on my laptop computer, proofread portions on my phone, and revised it on my iPad. My professional life reflects the sharp turn that new technologies are taking in the growing global economy. Although some critics argue that mobile media, social networking, and virtual environments "flatten" the world and affect communication, services, and job skills globally, this flattening process also greatly increases America's academic competition in a global society.[4] In recent years, the United States has fallen to number 12 in global rates of college completion;[5] a generation ago, the United States was at the top of these rates. Debates revolve around how we should get students prepared for, excited about, and matriculating into four-year colleges and globally competitive careers. And what place does teacher voice have within this discussion?

Too often, I've seen educational policy debates discredit the tacit, day-to-day experiences of teachers. In my eight years in the classroom, I seldom felt that the core relationship between teachers and students was what was most valued by the educational research and policy changes that flooded my school. Although the details of my own school's student-organized walkouts and police-sanctioned lockdowns (both detailed later) may differ from the experiences of other schools across the country, they represent the kinds of endemic disruptions and challenges that urban educators face. At the same time, the shortcomings of urban schooling that are documented by writers like Diane Ravitch and Jonathan Kozol end up reinforcing stereotypes about what happens *inside* schools like South Central High School. To use a problematic phrase, this book shares a perspective from "in the trenches" of public education. My emphasis is not on spotlighting the sensationalist images of school squalor run amuck but rather on illustrating powerful opportunities for shifting learning and teaching for the better. To use researcher and teacher Jeff Duncan-Andrade's phrase, "critical hope" drives the vision of paradigm-shifting change in schools that I articulate in this book.[6]

At the heart of this book are a series of basic questions. Considering how different youth culture is today than in the past, can mobile devices like student phones act as powerful tools for academic and critical learning? How are these new devices shifting the culture of school campuses, creating new forms of social participation, and shaping youth civic practices? And—perhaps most important—what roles should teachers, trust, and caring relationships play in an increasingly digitized school environment?

These questions instigated ongoing, in-depth research—working with students at South Central High School, dialoging with urban youth from other schools in Los Angeles, and conducting ethnographic fieldwork to understand and frame how young people use mobile media devices in their social lives at school.[7] Looking at the initial data from this research, this study led my students and me to develop in-class games that redefined what happens in a ninth-grade English classroom. Occasionally, our class activities landed me in my principal's office: mobile digital tools have shifted how individuals interact and communicate globally, but they were disruptive to traditional school rules at SCHS. The differences between how young people learn and how schools *expect* them to learn is staggering, and these changes are rooted in transformed culture, shifts in technology, and challenges to adult authority.

The New Landscape of Youth Engagement

One of the greatest shifts we've seen in how youth learn and interact with each other today has come as a result of our new "participatory culture." As described by media scholar Henry Jenkins and several of his colleagues, "participatory culture" is a general shift in culture from merely consuming media to also regularly producing it.[8] Making posts on social networks, uploading photos and videos, and participating in transmedia activities are all examples of these adjustments. When we watch reality television shows and respond to various hashtags or participate in a poll related to an evening's episode, we are engaging in some of the ways that mainstream media has adopted aspects of this new participatory culture.

In looking at what participatory culture means for educational contexts, Jenkins and his colleagues describe five aspects of this culture. In particular, a participatory culture is one

- With relatively low barriers to artistic expression and civic engagement
- With strong support for creating and sharing one's creations with others
- With some type of informal mentorship whereby what is known by the most experienced is passed along to novices
- Where members believe that their contributions matter
- Where members feel some degree of social connection with one another (at the least they care what other people think about what they have created).[9]

Youth engagement today—sharing and building Minecraft creations, uploading new music tracks for others to listen to and remix, playing e-sports (competitive video gaming), finding new youth celebrities via Tumblr and whatever the next social platform is—frequently relies on participatory culture. As implied by the list above, the possibilities of participatory culture often are ushered into mainstream society alongside advances in mobile technologies.

Mobile media have fundamentally transformed society. Much more than a telephone, today's mobile media devices connect most places around the globe to key networks, economies, and civic opportunities. These ubiquitous devices are often banned from classrooms and seen as problematic distractions. As a former English teacher with eight years of experience in an urban high school, I've seen how mobile media have transformed students' lives. Considering that most urban youth of color regularly use mobile media devices such as cell phones, I have spent a large portion of my teaching and researching career trying to understand how these devices and student predilections for gameplay can help students learn.

In my experience, the adoption of mobile devices at the turn of the century was swift and wide reaching. Over the eight years I spent as a teacher, I watched as students moved from tapping keys to swiping screens and from holding their devices in the air to search for a cellular signal to figuring out the continually changing Wi-Fi password for the building. From immigrants recently arriving from Central America to youth who spent their entire lives in urban Los Angeles, most of my students relied on mobile devices to connect with family and friends and to stay abreast of their complex social networks inside and outside of school. Aware of my own privilege as a middle-class adult, I was intrigued by the fact that nearly all of my students had their own mobile devices. The ubiquity of such expensive products was not lost on me as I entered this research.

At the same time, because they encourage "anywhere, anytime" learning, mobile devices signal a ripe area for policy investment in order to improve student achievement across socioeconomic levels.[10] Although mobile devices can offer significant opportunities within classrooms when properly situated in a wireless critical pedagogy, simply glorifying these devices is dangerous. The possibilities afforded by mobile devices are contingent on teacher support and the teacher-student relationship. As my own students helped me understand, bringing wireless devices into the academic realm of a classroom shifted these devices' context. They felt that the meaning of a phone immediately changes when brought into the adult-governed realm of a classroom. Without responding to and being aware of the contextual changes that mobile devices undergo when used within classrooms, their potential for learning ceases to move toward a sustainable reality.

Although a participatory culture is a key part of youth engagement, it frequently is seen as a source of distraction and discussed as a cause for alarm in schools.[11] However, actual empirical data about utilizing mobile media in schools is severely lacking. Schools across the globe have adopted tablets and mobile devices for students to use, but little research has been conducted on them in schools, and teacher preparation has not been focused on these tools. At the same time, teachers often have been discouraged from using student-owned phones in the classroom, even though current research suggests that most students—across socioeconomic lines—own these devices.[12] Further, recent research suggests that low-income mobile phone owners are disproportionately likely to access the Internet only from these devices. This means that they use these devices to access the larger civic world but are discouraged from using the devices during school hours.[13] Although educators have debated and tried to address a socioeconomic "digital divide," this debate is now largely in our past because most students can regularly access mobile technologies.[14]

Instead, today there is a continual "participation gap" between students who have multiple avenues to access and engage with new digital technologies and those who are limited to mobile devices (and occasionally dilapidated in-school computers).

In a 2001 article titled "Rethinking the Digital Divide," Jennifer Light examines the ideological assumptions historically behind the role of computer use in education.[15] She explains that the arguments surrounding the digital divide act as a frame for a "public problem" so that the digital divide becomes a way for the public to discuss larger systemic issues of inequality. I would argue that we are still discussing systemic issues of inequality through a digital frame—although today the frame is how we assume students use mobile devices both in and out of schools today. Light's analysis continues to extend, more than a decade and a half later, around issues of youth equity. Light explains that most policy reports related to the digital divide make two assumptions—that introducing computers mitigates inequality and that digital life "frees individuals from other social constraints."[16] The public discourse of these problems makes sweeping assumptions about the role that technology plays in the future. Light points out that this is an attempt to find a technological solution that redefines social inequality as a problem of access. Overemphasizing a hope that technology will solve school inequities avoids addressing underlying systemic problems.

By looking at the particular uses of technology in my school, I am hoping in this book to reorient some of the conversations our nation is having about the roles that should be played by participatory culture, student interest, and teacher preparation. This book does not presume that using technology and leveraging participatory culture in schools are quick fixes for the educational achievement gap or issues of civic development. However, in looking at mobile media and participatory culture in schools and my classroom, I offer this work as empirical data about the ways that these tools might increase equitable learning opportunities across the nation.

In a review of technology use in American classrooms since the 1920s, Larry Cuban writes that the general approach of teachers is to adhere to "constancy amidst change."[17] He notes that "Those who have tried to convince teachers to adopt technological innovations over the last century have discovered the durability of classroom pedagogy."[18] Overhead projectors, filmstrip and film projectors, video recorders, computers, and LCD (liquid crystal display) projectors have all found their way into traditional models of classroom instruction. However, considering how frequently students text, update, and generally *use* their mobile devices, classroom practice must recognize the digitally immersive nature of communication and literacy development that inculcates student identity formation everywhere *except* the classroom. Unlike projectors (overhead and LCD alike), teachers and administrators are no longer introducing the digital tools that now appear universally in schools. Instead, these tools are in the hands of students and no longer under the control of teachers.

Critically Connecting Learning

Each year, technology gets a little more advanced. It gets faster, smaller, and more necessary in work and educational contexts. Online social networks expand its reach. Digital devices get a boost of resolution and battery life. Devices can be worn on the wrist and can immerse us in three-dimensional worlds. They can print out handheld objects seemingly created from the ether.

Kids these days, we may think.

Simultaneously, this country spends a lot of energy trying to make schools better. But better at what? What is the purpose of schooling in the twenty-first century? When tangible "things" are printed out and household appliances are connected to the Internet, what are we connecting school learning to? What—we should be asking loudly—is the point?

As fancy as we think technology may be, efforts to integrate new technologies in schools in order to meet the needs of today's youth have been scattershot and heavy-handed. Each year schools and districts return to the well when it comes to school innovation: they invest in technology. Larry Cuban has spent decades researching these failed attempts, which he calls the "perennial paradox" of education and technology today.[19]

This may sound contradictory, but powerful learning with technology is never about the technology. Instead, I believe education needs to take a closer look at how teachers are being prepared to respond to and learn *with* the children in our classrooms. Technology can help facilitate classroom learning, but it will never be the panacea that politicians and school administrators hope it will be. This book looks at teachers and students as a foundation for powerful school-based learning, with technology augmenting the human potential in schools.

Rather than looking to technology for answers, I want to point to what a collective of researchers has recently called "connected learning."[20] Today's youth engage in a plethora of activities—leveraging interest-driven pathways for learning with peers and a global community through the use of technology—that can be described as connected learning. For instance, one of the most popular games that is being played globally today is Minecraft. As a digital sandbox that allows players to build to their hearts' content, raze their creations, and embark on traditional video-game combat, Minecraft is an expansive game that can mirror the nondigital creativity of Lego. New players of Minecraft can collaborate with peers within the game, discuss their designs, adapt those of others, gain and share expertise in a multitude of ways, and above all *make* things within Minecraft rather than have the content dictated to them. These possibilities then embody the six key principles of connected learning:

- Peer supported: Feedback is freely given and solicited.
- Interest-powered: Engagement aligns with participants' interests and passions.
- Academically oriented: Learning ties back to powerful, academic contexts of engagement.
- Production-centered: Things are *made* rather than simply consumed.
- Shared purpose: Others with similar interests create a robust ecosystem for learning.
- Openly networked: Resources are found in abundance across varied texts and modalities.

Looking at these six principles, it is easy to see how a digitally connected game like Minecraft can support connected learning. More important for educators to recognize is that connected learning *happens*. It doesn't require a teacher, a school, or a standardized test. It requires little more than for someone to be interested in something that others are also interested in. When educators can build on this context of learning, we shift what classroom engagement looks like.[21] In describing how connected learning functions today, Mizuko Ito and colleagues in *Connected Learning: An Agenda for Research and Design* highlight the framework's emphasis on equity:

Connected learning centers on an equity agenda of deploying new media to reach and enable youth who otherwise lack access to opportunity. It is not simply a "technique" for improving individual educational outcomes, but rather seeks to build communities and collective capacities for learning and opportunity.[22]

Although the above framework places connected learning as a contemporary form of engagement that youth engage in because of advances in technology, I think it is important to challenge two parts of this argument: (1) connected learning does not describe only youth-driven practices (adults engage in connected learning as well), and (2) connected learning does not always rely on new media.

I've written elsewhere about how teachers are already framing classroom practices as connected learning,[23] but throughout this book I highlight how connected learning is built on relationships. Although some technology is used during forays into establishing a "wireless" classroom in chapter 3, my emphasis is on how teachers and students build from a foundation of trust for significant changes in schools. I am hoping that you take away from the following chapters the fact that technology is never going to replace the powerful learning that takes place when teachers focus projects on work that students are passionate about. Technology can assist this work but cannot replace it wholesale. Despite connected learning's constant emphasis on digitally augmented contexts of learning—for example, in online video game communities or in the forums of knitting enthusiasts[24]—educators need to consider how the six principles of connected learning are manifested within their classrooms on a daily basis. And if schools struggle to locate these principles in their classrooms daily (as I suspect most presently are), then it is time to realign the work of teachers to meet how we know students are learning and are excited to continue doing so.

Before returning to the key research questions that unfold throughout the remainder of this book, I want to note that connected learning alone is not enough to reimagine what happens in classrooms today. The work we do with young people must also be *critically transformative*. Concerns that schools may not be relevant today have led a vocal group to argue against college attendance and against schools.[25] I have a more optimistic perspective on the value of schools as a space for socializing and guiding the interests of students. I see schools as spaces for consciousness-building and social transformation.

Toward Equitable Playing and Learning

Participatory changes, the possibilities of connected learning, and the ubiquity of mobile devices in schools are affecting how young people interact and learn, but can these changes lead to sustainable and powerful changes within classrooms? All too often, as an urban teacher, my school was beleaguered by one quick-fix solution after another. Entering this book's research project, I was not interested in a one-and-done approach to working with students. I wanted to get to the root of how technology is *already* affecting what's happening in schools and how the unspoken participatory culture could be supported without implementing expensive interventions or hiring consultants that lack school contextual knowledge. Further, such programs must respect the needs and interests of students in our classrooms. I wanted to explore how changes can support teachers—who are the most consistent presence in the lives of youth during "business hours" day in and day out. During my eight years as an English teacher, I had eight principals. With each successive principal, a new vision for reform at my school rolled through our professional development meetings. These reforms—new tests, new things that administrators required us to post on walls, new ways to measure—all pointed to a recurring lesson: our school was failing, and it was the fault of teachers. Or parents. Or the previous principal. Lots of fingers were pointed, but momentum around these reforms was often lacking.

And yet what if the reform we need is already here? What if the participatory culture and connected learning that captures the interests of students outside of schools is something that could be aligned and utilized within schools? I undertook this work because the topic means a lot to me. As a high school teacher, I shared stories of grief with colleagues about students' relentless use of mobile devices in my classroom. Hearing frustrated teachers from down the hall vow that they would confiscate student's devices and seeing students stealthily respond to or instigate conversations on their phones when they were supposed to be *learning,* I felt that participatory culture was secretly thriving and subtly adjusting how school time was used and understood. With confrontations between students and adults around using mobile devices, accessing the Internet, and confiscating devices, the realities in schools have been tricky, even as school districts move to one-to-one initiatives allowing all students access to school-sanctioned devices.

In the following chapters, I look at how I've used games and student mobile devices in my own classrooms to focus youth engagement. Implementing low-cost structural changes—such as using student-owned phones and principles of gameplay that reflect participatory culture—can quickly recalibrate schools to reflect the society we are preparing kids for today. As current research identifies a digital participation gap that largely aligns with the U.S. education achievement gap, analyzing the role played by digital culture and literacies is nothing less than an issue of equity and civic justice.[26] One challenge that I faced in designing this research is that although numerous studies point to the learning potential that new media pose in the changing landscape of education, little research focuses specifically on formal learning environments like schools and classrooms.[27] As such, this book looks to the unique challenges within SCHS that teachers face in incorporating new learning practices. Some of the challenges that I've noticed and experienced during the time that I conducted the research for this book include working within existing school rules, addressing socioeconomic factors related to utilizing technology in underserved urban communities, and preparing teachers for the new forms of work that come with the use of these new tools. SCHS is not perfect. No school is. I offer the blemishes of this campus in order to look at real students in a real public school.

Just handing a device to a student is a foolhardy approach to academic change in schools. Researchers like Larry Cuban have conducted powerful research about the lack of classroom changes that have happened with new technology advances over decades of attempts at this point.[28] Likewise, my own research, particularly with my colleague Thomas M. Philip, has explored how assumptions that technology will "fix" today's iGeneration get in the way of actual classroom innovations.[29] Instead, the shifts in culture tied to mobile-device usage and online activity among youth and adults alike change what in-class engagement looks like. For example, in the twenty-first century, it's no longer enough to look at only youth and teacher communication during the hour or two that a student is in my class. Text messaging, Facebook, online grading systems, and teacher-focused social networks create sustained forms of communication long after students leave my classroom. (Students in my college classes who are impatient for me to respond to e-mails have learned that sending me a tweet is likely to get the quickest response.) The participatory nature of online resources like Facebook, YouTube, Skype, and *Wikipedia* reflects how individuals interact with, learn from, and critique media products. These differences require adjustments in classroom interaction and learning practices.

As I looked at the challenges faced by the school I taught at while writing and researching this book, I questioned the college-ready vision of students that the country's education system outlines. In particular, traditional "college and career readiness" preparation that happens in high schools lacks a contemporary articulation of how America's youth will interact, engage, and shape the public world around them. As I looked at schools across the country and at the challenges in my own classroom, I asked myself a few questions: What is the *civic purpose* of schooling when it is narrowed simply to a means of attaining passing results on high-stakes tests and completing college? Related, what roles should be played by imagination, innovation, and ingenuity in classrooms today? *When are schools fun*? Outside of schools, we have a strong understanding of the robust ways that youth are "hanging out, messing around, and geeking out" in intellectually rich contexts.[30] But in schools? Well-intentioned pacing plans, interventions, and assessments often get in the way of experimenting with teacher innovations and developing meaningful relationships with students.

I raise the questions and challenges in this book while acknowledging that education in the United States continues to remain stratified across race and class. Numerous reports note that the gap in economic inequality in adults is widening over time and is intrinsically linked to academic achievement.[31] And although large, sweeping studies have focused on the systemic economic inequality that affects American life,[32] efforts to describe how to instruct and teach about this omnipresent issue seem mired in focusing on singular lessons.[33] This study looks at how in-class instruction related to local civic issues could help mitigate issues of academic and economic inequality.

In order for *every* student to be engaged in meaningful academic and civic development, today's public school children need to be immersed in learning environments that engage the multifaceted nature of learning today. These environments, mediated by mobile media and gameplay, could level learning opportunity disparities across race and class in America's public schooling sector. At the same time, educators need to gain a more nuanced understanding of what is happening within classrooms and schools with mobile media devices. As a teacher, I often assumed students were busily communicating, producing texts, and staying abreast of social changes within peer networks all while sitting within my classroom. Parts of my assumptions proved correct, and the extensive time I spent talking with and observing students illuminated for me the robust set of skills and attitudes students today possess in regards to mobile-device usage in schools.

Where We're Headed

The following five chapters of this book present a year in the life of the students and teachers at South Central High School. From an initial tour of campus life to day-to-day instruction in my third-period ninth-grade English class, this book represents a snapshot of public education over one year at an urban high school. During this year, I spent time teaching in my classroom as well as talking with kids from across the school in order to investigate the effects of participatory culture and technology. Instead of looking to other sources for improving schools, I examined the ways the technology already available and utilized by students could be leveraged for critical thought and civic participation within a classroom.

In addition to focusing on what happens within my own classroom and the relationships forged with students, this book begins by looking at mobile media use among all students at SCHS. Chapter 2 details the assumptions, practices, and challenges that students and teachers face while using mobile media at school. Although recent reports highlight the increased use of mobile media by youth, actual data about in-school use is lacking.[34] Significant studies focus on how students are currently engaging in digital media through mobile devices and writing practices, but most of such work examines the informal learning environments that students are working within.[35] This book helps close the gap in our understanding of youth learning today. Through focused qualitative research, I looked at what students were doing between the hours of 7 a.m. and 3 p.m. at my school, regardless of what school policies dictated that they should be doing. I also include an analysis of student and adult conflicts in day-to-day interactions, such as when a student is late to class.

The remainder of this book utilizes these initial campuswide findings to develop a framework for instruction in my own classroom. To try to reframe classroom instruction, I handed out a class set of iPods and developed an alternate reality game (ARG) for my students in order to shift the contexts of in-class learning. I look at my initial experiences (including blunders) with mobile technology in the classroom throughout chapter 3 and follow this with a look at the game design that occurred in my class. Chapter 4 focuses on Ask Anansi, a game I created about storytelling and revealing the hidden narratives that guide our understanding of power and control. With this game, I designed and implemented a formal curricular project that engendered student media practices and elements of gameplay for a ninth-grade English classroom. This chapter details the nuances of the teaching and pedagogy of this gameplay and curriculum and also brokers a larger conversation about the possibilities of games, mobile media devices, and learning in today's urban classroom. Finally, in chapter 5, I wrap up my analysis with a look at trust, changes in student leadership, and the limitations of expecting too much from technology.

Throughout this project, I collected qualitative data, including daily fieldnotes, interviews, and student-produced texts. Student work samples were gathered and measured, and I audio recorded class instruction for a close discourse analysis. Appendix A is a detailed account of my methodology for data collection and analysis.

Since completing this research, I've shared some of the findings from this study in publications aimed at educators and the broader educational research community.[36] However, considering the excruciatingly slow pace at which schools have changed, this book is written to invite a larger audience of readers to consider the possibilities in schools and reimagine what learning can look like as we dive further and further into the twenty-first century while still relying on archaic pedagogy.

A Note about Names

To protect the identity of the students who collaborated with me on this research, I have renamed our school South Central High School—to signal the general geographic community in which this school and this study are immersed and also to validate the counternarrative of cultural prosperity that persists in urban Los Angeles. Historically, films, music, and news headlines have depicted South Central Los Angeles with images of poverty, violence, and squalor. Mainstream media and the governing agencies of Los Angeles now refer to the community as "South Los Angeles," and this change was made official in 2003 by the Los Angeles City Council.[37] This action can be seen as an effort to erase a cultural past of uprisings, resistance, and negative press through a superficial renaming of the community. However, despite the flooding of "South Los Angeles" messaging in media, I have never heard any of my students refer to this community as anything but South Central.

Related, the name's stereotypical connotations of violence, gang activity, and generalized danger are made manifest through new media tools like search engines. One year my class critically analyzed Google image search results for popular professions like "doctor" and "lawyer." For the most part, the search results didn't surprise. Predominantly white, male faces showed up as the top results. As a class, we talked about what the search findings represented and why they didn't reflect our class and community demographics. The lesson was a place to continue our application of critical terminology like *hegemony* and *counternarrative* and to think about the mechanisms that could change the results of this search in the future.

One of my students returned from our Thanksgiving school break with an exciting class activity. Demonstrating for the class, my student typed into Google image search "Beverly Hills." He said he noticed all of the clean streets and smiling white people. Next, he typed into Google image search "South Central Los Angeles." The contrast was striking—power lines, fast food, gangs, police making arrests. As a class, we discussed what stories are being told about these two communities. What is being left out and why? We continued to explore the "dual cities" in Los Angeles and the ways we could remold the story being told about our school.[38]

Finally, the names of students, teachers, and school officials in this study have been changed. Their names still account for traditional gender conventions (a student who identifies as female in the study is given a traditionally female name). As an English teacher at heart, I have given names that signal specific literary themes and concepts. For instance, one student in class is named Dante in a nod to the journey of the poet and protagonist of *The Divine Comedy.* Although these names are ancillary to the principle research questions explored herein, the names invite a playful dialogue between myself as researcher and curator and you, the work's reader. All of these names are derived from works of literature that I discussed with students in my classroom. Conversations with my classroom community about works by Miguel de Cervantes, William Shakespeare, Ralph Ellison, and Julia Alvarez have significantly enriched my passion to teach and my understanding of youth cultural and academic practices within the South Central communities to which my students belong.

1 Walls, Silence, and Resistance: A Tour of South Central High School

Getting in and Getting out: Welcome to South Central High

The first time I visited South Central High School, it was a Thursday morning in 2004, and at least a dozen police cars were parked on the street in front of the school. They were a black and white barrier, covering more than half the red-curbed block. In addition, the local "ghetto bird" (the students' name for a police helicopter) was on a not infrequent search for a criminal in the school's surrounding streets. The feeling was oppressive. It was hard not to feel suspect, under someone else's control, and powerless when walking into the school. Later in the day, I learned that the police were at this location for a monthly meeting held in the school's cafeteria. But rather than making the school feel safer when I entered it, their presence made me feel on edge in the school. In and around the classrooms of SCHS, security at the school and around the neighborhood is always perceived as heightened.

From the outside, SCHS looks like a typical West Coast open-designed campus (aside from the constant presence of police at or around the school). People enter and exit through the central administrative building during the day. The campus takes up an entire block and is comprised of decades-old buildings and aging "bungalows" that once stood as temporary classrooms. I never heard of any plans for phasing out these classrooms.

The SCHS campus is surrounded by a twelve-foot-high metal fence that is topped with curving spikes facing outward. Friends and family members often pass food between the bars to students during lunch breaks, which conveys a prison-like feeling to school life at SCHS.

Getting into and out of SCHS can feel just as daunting as this image suggests. A security guard stands by the front door, which usually is only cracked open and often is shut and locked. Campus visitors need to knock loudly and hope someone is on the other side of the door waiting to receive them. Moving past this door, visitors need to show identification and sign the register before they are directed to the main office, where they will be assigned an escort to accompany them to whatever office or classroom they may be headed toward. This multistep process can feel confusing and intimidating. Most of the parents I've worked with at SCHS are native Spanish speakers, and the bustle of language from one door to another (even though some of these positions are staffed by bilingual employees) can deter them from being involved and engaged in the lives of their children.

Students and teachers don't necessarily have it any easier. (I detail the rigid processes that students go through to leave the school on a daily basis later in this chapter.) As a new teacher in my early twenties, I found it difficult to enter the school or leave during my planning period, depending on who was at watch at the front of the school—even though I regularly showed up to work wearing professional business attire. (I had internalized at an early age the notion that I should dress up if I am going to meet somebody important, and I knew that my students were the most important people I would encounter during my work day.) The school eventually instituted a uniform policy, and I switched between collared shirts and ties and the uniformed attire that students wore. Other than a penchant for wearing worn-down Converse low-top sneakers, my clothing usually didn't match that of my students. And yet I regularly was stopped while heading for the door and asked where my pass was, why I was out of class, and where I was going—despite the lanyard with my name and position that hung from my neck. Even though I was at the school every day and saw the same security guards, this was a daily rigamarole that had to be endured whenever I needed to go to my car, buy lunch, or be off campus for a short time. Such small inconveniences followed me throughout the campus for most of my teaching career: security asked why I was out of class, why I was using the teachers' restrooms, and why I was parking in the teachers' parking lot. Being Latino and walking with a backpack may not have helped. Although it is easy for me to complain about the heightened security, this thoroughness was intended to increase safety for students and adults on campus. But such experiences illustrated for me the daily indignities that teachers and students experienced because of the structures that governed SCHS.

Educational researchers Daniel Solorzano, Miguel Ceja, and Tara Yosso have looked at the role of "racial microaggressions" in educational contexts.[1] Described as "subtle insults (verbal, nonverbal, and/or visual) directed toward people of color, often automatically or unconsciously,"[2] microaggressions are a constant factor for students at SCHS and are pervasive in how the students quoted in this book view and describe the world. From the language that security guards use to address students to the wariness that police show around them, microaggressions are often paired with adult skepticism about students. In addition to shaping the daily experiences of students, microaggressions also demoralize teachers.

As an eager new teacher on my first day visiting the school, I felt intimidated by the appearance of a decrepit school that seemed to be surrounded by police vehicles. After spending years within the school, however, the space felt fulfilling and hopeful to me. Although SCHS can offer myriad bleak tales of school failure and underperformance, at the heart of the campus is a thriving and beautiful hunger for knowledge and education. Many of my students were brilliant, and my teaching colleagues were dedicated and creative in their pedagogical approaches. In this chapter, we'll explore campus life, key policies that affected classroom learning, and the academic challenges that led me to confiscate mobile phones in my classroom, seeking out good—humanizing—interactions with my students.

Further Exploring SCHS

Beyond the imposing metal bars that keep SCHS students in and trespassers out is the city of Los Angeles and its traffic-laden freeways. SCHS is one of the oldest public high schools in the city of Los Angeles and celebrated its hundred-year anniversary just prior to the year captured in this book. As generations of students matriculated through (or dropped out of) the school, new buildings were added to the growing campus, coats of paint were added, and older structures were converted for new purposes—all in an effort to keep an archaic campus up to speed. The school's book storeroom, for example, started out as an auto shop.

Likewise, a former shop classroom housed two different classes at the same time, leading to occasional excessive noise.

Figure 1.1
An auto shop converted into a book storeroom

As the enrollment at SCHS began to exceed the capacity of its buildings, temporary classrooms were installed on campus. These temporary buildings were euphemistically called "bungalows," and they were not "temporary." By the time I arrived, these buildings had aged along with the rest of the school, and there were no plans for them to be removed or replaced with something more solid. These were some of the first classrooms I taught in at SCHS. In my first year, my students used a broomstick to measure the depth of a hole in my floor and found that it went down four feet to the earth below.

Figure 1.2
An auto shop converted into a book storeroom

Along with the physical changes, the century-old campus has undergone significant challenges and demographic shifts in the last few years. One of the main changes was that students were funneled into and out of the school twelve months of the year. With a student population of approximately 3,400 students at the time, South Central High School was one of the largest schools in the city and one of the few that operated on a year-round three-track schedule. To accommodate all enrolled students, a third of students are always "off track" and not scheduled in any classes. To deal with overcrowding at SCHS and other schools throughout the city, the Los Angeles Unified School District moved these schools to year-round

Figure 1.3
The view from one class into another

Table 1.1

	A track	B track	C track
July–August	Break	In session	In session
September–October	In session	Break	In session
November–December	In session	In session	Break
January–February	Break	In session	In session
March–April	In session	Break	In session
May–June	In session	In session	Break

schedules that meant that other than for a two-week holiday in December, the school was always operational.

My students were on the B track of the year-round calendar. Our school year started a day or two after July 4 and ended the last week of June. During those twelve months of our academic year, we had two two-month breaks. For educators, this schedule was wonderful. Teachers could travel during off-peak seasons and have long rests between four-month-long stints of teaching. Further, instead of summer school (B track's school year begins in the middle of summer), SCHS offers students "intersession courses" twice a year—six-week-long courses for credit recovery and occasional electives.

During intersession one year, one of my students came into a film class I was teaching visibly upset because he'd been issued a truancy ticket for being outside of school during school hours. The municipal law states that minors under the age of eighteen cannot be outside "loitering" between the hours of 8:30 a.m. and 1:30 p.m. Even though B track was in intersession (the equivalent of summer break), the problem was that our class started at 10 a.m. The next day, I saw a police officer who was issuing a ticket to another of my students. He said that students either needed to be accompanied by an adult (which is an unrealistic expectation because their parents are working) or arrive at school by 8:30 and wait for the class to begin. This response shouldn't have surprised me. It fit together with the microaggressions and other policies facing youth of color who actively seek learning opportunities. To comply with this attendance policy, my students met me across the street from the school each day, and we walked in together—a coordination process that ate into our instructional time. Some of my best students missed class to go to court to fight truancy tickets they received while trying to go to school voluntarily. My students were not surprised by these events. They understood that structural barriers stand in the way of their getting the educational experiences that are the right of all young people living in Los Angeles.

Looking again at the literature that correlates student performance with socioeconomic background, I can't help but connect the truancy example with the school's racial makeup. The demographics of the school's student population mirrors those of its surrounding community—83 percent Latino, 15 percent black, 2 percent multiracial, with an English language learner (ELL) group that makes up 39 percent of the students. Additionally, 87 percent of the students receive free or reduced-price lunch.[3] Free and reduced-priced lunch is a useful means of measuring household income. Looking at the surrounding elementary schools that eventually feed into SCHS, the number of students receiving free lunch is closer to 100 percent. The lower percentage at SCHS is not because families become wealthier as students go into high school but because some teenagers don't turn in the required paperwork.

The quantitative data about SCHS can reinforce perceptions built by the rough-around-the-edges appearance of the campus. During the years that I taught there, the school's dropout rate was always above 60 percent.[4] The data collected on students at SCHS (demographics, aggregate academic performance, and pathways after they leave the school) by and large frame SCHS as an urban school that—to use the language of the popular documentary *Waiting for Superman*—is a "dropout factory."[5]

How, exactly, does a school become an "urban" school?

More than a century ago, when Los Angeles was beginning to be developed, there was no city blueprint declaring that here would be a ghetto and there would be a suburb. A West Coast equivalent of New York's Robert Moses did not proclaim that South Central would one day become a stereotypical space for talking about dropouts, gangsters, and localized violence. And yet South Central High School and some nearby schools are in one of the most economically impoverished areas of the city and academically are among the lowest-performing schools in the city. Educational researchers and sociologists have produced significant research drawing connections between socioeconomic levels of youth and their academic performance and postsecondary choices.

For several years, my eleventh graders and I collaborated on researching and sharing the history of South Central Los Angeles. This work included conducting interviews with community residents and family members, diving into local history archives, and looking at media artifacts from across the history of South Central (often visiting the Southern California Library just a few blocks south of the school). As one aspect of this work, we gathered around a small closet in the school's library and looked through the near-complete collection of yearbooks from the school's opening through the present day. The activity illustrated that the school began with an all-white student body in the early twentieth century, shifted in the 1970s and 1980s to a primarily African American student population, and shifted again to a primarily Latino population. Along with making comical asides about changing fashion trends and hairdos over the decades, students saw a narrative of historical white flight as local industry jobs disappeared decades ago.

In addition to the dwindling workplace options for students in the neighborhoods around the school, the school struggled to offer powerful college counseling for the increasingly large student population. While I was teaching at SCHS, the school had one college counselor for the entire campus. In the year of this study, only 35 percent of students graduated from South Central High School. Of this group of students, 14 percent completed the required courses to be eligible to apply to most four-year universities.[6] Too often, rhetoric in education debates is about decreasing the dropout rate, but it ignores the fact that students at a school like SCHS can graduate from high school and still not be eligible for postsecondary options other than increasingly overcrowded community colleges.

A Brief Note about the Police

Before continuing this tour of the campus, I want to briefly address language I use throughout this chapter when discussing the police. Nearly a decade after my first visit to the SCHS campus and its police-car facade, I was house hunting with my future wife for a home in Fort Collins, Colorado. My mother was with us on the day we traveled with a real estate agent from one housing development to another across the quiet city. One house seemed like a contender: the layout felt right, the backyard was reinforced enough to thwart our dog's constant schemes to escape, and there was ample room for the books a professor and librarian haul with them from one abode to another. However, when I looked outside, I saw three different police cars parked at adjoining homes. Our real estate agent pointed out that many officers lived here because it was close to a local station. As close-minded as it may sound, I told the real estate agent I couldn't live surrounded by police officers. I wouldn't feel right. My mother raised a skeptical brow about what I might be planning to do in our new home.

I grew up in a suburb of San Diego and was raised to turn to the police if I was ever in trouble. They were people I could expect to help me. I also have never been harassed by the police at or around SCHS. But I have seen several of my students escorted out of the school in handcuffs, dropped to the floor with a controlling knee digging into their backs, and sweating uncomfortably in the back of patrol cars. Some of my students have been stopped on their way to campus, and others have received truancy tickets for coming to campus on the days their official track was not in session. Law officers have important responsibilities in our communities and are exposed to physical dangers that most of us never encounter.

But my years teaching in SCHS showed me that the relationship I had with police officers and the law through my middle-class upbringing was drastically different from that of my students and the nearly entirely black and brown community of South Central Los Angeles. Some communities are criminalized and treated with very different assumptions about who lives there and what the value of "their bodies" is (to quote Ta-Nehisi Coates) than other communities are.[7] However, recent scholarly texts like Michelle Alexander's *The New Jim Crow: Mass Incarceration in the Age of Colorblindness*[8] and Matthew Desmond's *Evicted: Poverty and Profit in the American City*[9] explore the overwhelmingly negative effects that a constant police presence, activities of police, drug policies, and the criminal justice system have on the lives of youth of color. In this book, I look at the roles played by police, campus security, and adult control of the school's campus.

"Ladies and Gentlemen"

In preparing to explore in-class uses of technology and the ways that digital tools could affect classroom learning, I spent eight weeks sitting in the lunch area of SCHS after lunch and after school. Although most people might not think that instruction takes place in this kind of space, I was able to observe the important lessons that students learn daily in their interactions with ever-fluctuating campus policies.

On the two days of my week not spent in student clubs or teacher union meetings, I watched students trickle back to class or home as soon as the school bell rang. As they headed into classes in the morning and through the gates that led them home at the end of the day, students created migratory patterns as they moved toward and away from key spaces for socialization. As a teacher, I was usually in my classroom by the time the bell rang to signal that lunch had ended, and I often saw students slide into class just seconds before they were considered late. Although I tended to not enforce the policy, students who were not inside a classroom at the moment the bell rings were required to return to the "tardy line" and go through a bureaucratic process with an escalating level of punishment before going back to class.[10] After I spent time outside of my class and observed what happened on the other end of the lunch bell, I better understood the lessons students learned when they are tardy and not in my class.

Perhaps the most obvious fact about teenagers—some kids are frequently late—I learned long before observing the tardy line. There are myriad reasons underlying this fact—and many are shaped by socioeconomic factors like taking public buses, supporting other siblings, and working part-time and full-time jobs. Some other students just "can't get right" (to quote one of my teacher education professors, Jeff Duncan-Andrade).[11] Teachers know which students somehow slouch into their classrooms a few minutes late every day, no matter what period.

Junior, an eleventh grader in my class in my first year of teaching, was my first student who was chronically late. He was also the first student I felt I failed to connect with and then failed to keep track of after he dropped out. This was despite my varied efforts to motivate, reprimand, and cajole Junior to show up on time (or at all). As I reflected on my blog years later,[12] I often used my conference period to cross the street to the gas station where I knew I would find Junior loitering and tell him he was missed in class. I found the phone number for the pay phone at this gas station and called that number when he wasn't in class, and he learned to hang up when he realized his English teacher was hounding him. And soon his name was dropped from my class roster.

As part of a school system, I spent a lot of time trying to *control* Junior rather than build the meaningful relationship that might have persuaded him to be present in my classroom. Years later, when I undertook this study, I better understood how my adult authority contributed to Junior's outlook. A student like Junior spends a lot of time at SCHS being ushered toward, standing in, and being "processed" at the end of a tardy line. Often, the interactions between students and adults in this tardy line turn into microcosms of disruptive aggression. Students get frustrated by how they are treated, and adults get frustrated at the lack of obedience displayed by unruly adolescents.

Despite attempts to improve the learning environment at South Central High, students I observed in the tardy line regularly complained about their treatment by teachers, administrators, and police at the school. The line for students to receive tardy passes before going to classes occasionally bulged with hundreds of students.

Like a psychotic Disneyland line, this seemed like the unhappiest place on earth. The students mobbed and sprawled in what only somewhat suggested the definition of a line. I saw occasional gang-related skirmishes, particularly when a new "track" of students cycled onto the campus every two months. Trying to maintain adult order over the transitions as new students funneled into the school mirrored the moment-to-moment transition of trying to corral thousands of students into classes in a seven-minute passing period. These efforts took the same shape and were primarily a set of commands barked at masses of students:

"Ladies and gentlemen, let's go!"

"Get to class, ladies and gentlemen."

"Keep it moving, ladies and gentlemen."

The adults who attempted to move kids to classes from the main quad frequently relied on standard repertoires of disciplinary practice, like jazz musicians improvising from a set routine of chord and rhythm changes.[13] Their daily efforts to compel students to go to class or to leave campus after school had a typical shape: an assistant principal or campus security guard spoke loudly in the direction of groups of students. The nuances of this practice—volume, distance, and diction—are important to consider. How loud, from how far, and with what words adults spoke to and *at* students illustrates the relationships between these two groups and their respective power on campus. Consistently, the adults used the same phrases in addressing the students of South Central High School.

One observation that I recorded exemplifies a typical encounter between students and adults. Some students were sitting at tables after lunch. No adults seemed to be patrolling the area, but from the south end of the lunch tables, I heard, "Gentlemen, let's go, let's go, let's go." Mr. Panza, the assistant principal for discipline, was wearing a leather jacket as he briskly walked through the area, carrying a handheld public address system. He seemed to be speaking to the group collectively and not to a specific student.

The words amplified from Mr. Panza's PA—"Gentlemen," "Ladies," and "Let's go"—were among the phrases most frequently used by Mr. Panza and other adults and were part of his personal disciplinary vocabulary. Initial observations of these language choices suggest that Mr. Panza and the other adults were engaging in a businesslike professionalism and were showing respect toward students. However, looking closer and seeing these exchanges again and again, adult efforts actually appeared to dehumanize these students.

During this study, I recorded and transcribed seventy different adult interactions with students that were close enough or loud enough for me to document. (Other incidents, I observed from a distance but, unless a bull-horn and PA system were used, could not discern the words exchanged.) Adults said "Let's go" 62.9 percent of the time and the words "ladies" and "gentlemen" nearly half the time (17.1 percent and 25.7 percent of the total incidents, respectively).

On the one hand, "ladies" and "gentlemen" were phrases typically used for respectful deference. However, within the school community, the language replaces individual students' identities. Instead of saying "Julio" or "Veronica," for example, students are referred to, en masse, as merely "gentlemen" and as "ladies." For example, I regularly noticed Janie talking with her friends. I observed eight instances of Janie being in the lunch area after the bell had rung. Often, an adult walked over to the group of friends, and—even though she is regularly seen in the area when the tardy bell is about to ring—addressed her group as "ladies." The adults never referred to her as "Janie" or even mentioned her constant presence in the lunch area, which to me felt like a lack of concern for understanding who the students were. Yes, these adults were interacting with hundreds of students on a daily basis, but part of humanizing education is about acknowledging, seeing, and *knowing* the individuals we interact with. In the following chapters, as I look at how learning was best supported in my classroom, the point I return to again and again is the need for adults to provide the space for trusting and caring relationships in schools. In place of humanizing practices, adults in the lunch area invoked formal rhetoric that—although useful for quickly addressing crowds of students—suggests that the adults don't really know student names or want to learn them.

There is a school of thought that argues that adults should forcefully move students into classes and off campus quickly after school. This approach can be seen as "tough love" that will help students get to where they need to be. The politics of words is largely invisible. After four years as students at SCHS (and likely experiencing similar language in elementary and middle school), the students I interacted with were treated with distance and a lack of trust. I question how the years of such distanced and clinically amplified language shaped how students see the world and their role within it. My own middle-class schooling experience was one in which such practices were not typical. I was occasionally late to class and sent to get a tardy slip or serve detention, but such small self-afflicted inconveniences were carried out by people who knew my name and likely cared about me.[14]

The after-lunch corralling process was replicated after school as well, when this effort pushed students away. Perhaps unlike your own memories of lingering after school to visit with friends and plan the rest of the day, SCHS adults tried to get student to vacate the campus as quickly as possible. The sooner students left, the better. Although the threat of a tardy line did not loom in the afternoon, the same sweeping commands were issued by adults to funnel students through the exit gates. There were few exit gates (so that adults could limit and control strangers entering the campus), but this meant that students regularly climbed over the campus fences to skip class, get lunch across the street, or avoid taking the extra steps to the official exits.

At the end of lunch, bullhorns and walkie-talkies sent adult voices over a sea of students. Even though there were many positive interactions between youth and adults at SCHS, the confrontation of policy and discipline after lunch and after school functions as a microcosm of frustration and dehumanization.

Michel Foucault describes the role of surveillance and control within prison systems: "The prison cannot fail to produce delinquents. It does so by the very type of existence that it imposes on its inmates."[15] However, my observations in these lunch areas illustrate student countermeasures and actions that emphasized defiance, subversion, and *sousveillance* to preserve student power. Students were able to ignore, hide, or otherwise resist the language of adults in these spaces, but such moments fizzled out over time. After students were funneled into the tardy line or the school was vacated, there were too few spaces for students to hide on campus before adults sent them to the enforced locations of the school.

If schools are the places where young people learn how to participate in adult society, then I am concerned about the lessons they learn through their interactions with adults at SCHS and the ways that processes of dehumanization at school reflect socioeconomic patterns in society. The cat and mouse game of authority and power enacted at the school several times each day shapes how students see the world around them. Policies and the ways they are enacted affect what students understand the purpose of school to be.

Campus Life and the Impossibility of a Policy Inventory

Many educational research books and articles have tried to delineate the various policies, agencies, and offices that affect how schools operate, what students experience, and what kinds of professional expectations are placed on teachers.[16] Often their Venn diagrams and flowcharts become a complicated soup of arrows, circles, and too many words written in too small a font. At the same time, the processes of ushering students through the hallways via distancing language are not an accident. The factors that influence what happens at a school are complicated.

Rather than attempt a partial overview of the many policies affecting students and teachers at SCHS, I want to continue this tour of SCHS by looking at a few key areas of educational policy that most directly affected my students and their experiences in my classroom. One of the more forward-thinking principals during my years at SCHS once attempted to inventory all of the extracurricular programs and clubs available to students so that students could browse this catalog and take advantage of the many choices that were available to them at the school. However, that principal left the school before the project was finished, so it became another microcosm of complex and intersecting choices for students to sort through on their own. Policy efforts have real implications on how students make choices and on how funding for their educational experiences is allocated. If SCHS cannot execute a plan to list all of the positive extracurricular opportunities available for students, how can the school's policies be implemented fairly and effectively, day in and day out, in the classrooms of thousands of students at the school?

Compliance Gestapos

In 2000, a class action lawsuit was filed on behalf of 100 students against the state of California claiming that the state and its agencies "failed to provide public school students with equal access to instructional materials, safe and decent school facilities, and qualified teachers."[17] The case of *Eliezer Williams et al. v. State of California et al.* was settled in 2004, just as I was taking charge of my first classroom. The case directly affected the day-to-day experiences of my students. The *Williams* case settlement required the state to allot nearly $1 billion to provide materials, improve facilities, and make other local improvements to underserved California schools.

When the settlement was announced, it felt like a windfall for schools serving historically disadvantaged communities, like SCHS. Learning spaces would get better, and books would always be available, which to me as a new teacher felt like the insurance I needed to help my classes prepare for success. Moreover, the state would send auditors to campuses to check on conditions and access to materials throughout each school year to make sure funding was spent as the lawsuit intended.

These changes did not happen.

Each semester, the distribution of textbooks felt lackadaisical until word was received that the *Williams* auditors were coming to school. Then there was a whirlwind of rushing textbooks into every classroom. When a man or woman came into class with a clipboard, students were prepared for the questions that they were asked:

"Do you have a textbook in this class?" "Yes."

"Where is it?" "Over there."

"Can you take it home?" "Yes."

This was the extent of how the *Williams* settlement supported learning in my experience as a teacher.

My class often centered on reading novels and plays—J. D. Salinger's *The Catcher in the Rye*, Shakespeare's *Romeo and Juliet*, and Julia Alvarez's *In the Time of the Butterflies*. As I interpreted the *Williams* settlement, it required every student to have access to the class text (not necessarily a textbook) in every class. However, an assistant principal often checked to make sure my students had large, expensive textbooks merely so that they could check off a box on a list. Every year, class time was disrupted to fulfill the district's compliance needs. Then interim Los Angeles Unified School District superintendent Ramon Cortines once described to me the layers of bureaucracy that get in the way of positive teaching and learning at schools, calling some district agents "compliance gestapos." He said that in this context "teachers should come first." Ultimately, the *Williams* settlement illustrated for me, as a new teacher, how a well-intended policy can interfere with classroom learning and result in wasteful in-school spending. Although I used the annual interruptions to explain to my students why we were issued textbooks and discuss the role that students played in setting legal precedents for educational equity, the entire experience shaped how students saw their time valued within schools.

Standards and Testing

On the other end of the spectrum of public awareness, educational standards and high-stakes testing are perhaps the most regularly reported on area of U.S. education today. Although most readers can remember how they were evaluated in schools and how tests shaped their entrance into professional careers, the standardization and assessment that have been institutionalized over the past two decades are probably starkly different than what you may have experienced. The general structure of classes, the periods allotted for learning, and some of the instruction in these classes look the same. However, the demands on teachers, the assumptions of what students *should* be learning, and the ways we measure this learning continue to be transformed from year to year. After George W. Bush signed the No Child Left Behind Act on January 8, 2002, an emphasis was placed on students' performance on exams (and later, a trickling upward of these results on teachers and on teacher education programs), and this has shaped the kinds of knowledge and experience that are happening in schools today. When I was teaching at SCHS, the school was still using the California State Board of Education's Content Standards. These were replaced a year after this study when the state (along with forty-two other states and U.S. territories) adopted the Common Core State Standards.[18]

Before getting into how standards are implemented and affect student learning and the teaching profession, I first consider what these standards mean. Typically, when I ask the preservice teachers I work with at my university to consider the word *standard*, they develop a definition that describes a specific, measured level of their instruction. This standard is what *all* teachers are working toward. It is both a ceiling and a bare minimum in this sense. Although this narrow definition defines what should happen in a classroom, it also limits learning experiences. Standards loosely address cultural and language diversity but typically do not reflect the lived experiences, interests, and societal shifts through which students see the world around them.

At the same time, I ask my preservice teachers to remember that all of us implicitly carry our own standards for the world around us. In this sense, I ask my students the following questions: What are the personal standards to which your own beliefs adhere? What beliefs about education, justice, and engagement do you feel should be reflected in everything you teach? Such questions reveal that standards like the Common Core are only one aspect of a series of protocols that are (or should be) standardized in our classrooms.

Finally, there is a third definition of *standard* that is less familiar in the teaching realm but immediately familiar in other contexts. In the jazz world, standards are the song charts that musicians need to know inside and out if they plan on making a career of live performance. Groups run through these charts, taking turns soloing and riffing off of one another. The standard is the backbone of the working musician. I think this is closer to the kind of space that teachers inhabited for decades prior to the current educational epoch of high-stakes testing. Teaching once was about honing a repertoire of proven practices that could be adapted based on the needs in a particular classroom.

And although I see much of my own work as an extended riff on the standards of pedagogy that I've practiced during my years as a teacher, I want to consider the kinds of melodies and mundane compositions that emerge from the policy standards driving classroom education today. South Central High School was designated as a school in "Program Improvement" for the entire time that I taught there, based on its students' performance results on California standardized tests. As a result of evaluation measures from No Child Left Behind, Program Improvement "places the school in a continuum of structured interventions designed to help identify, analyze and address barriers to student achievement."[19] However, Program Improvement has established specific interventions for only a five-year period, attempting to ensure that no school is designated as part of Program Improvement longer than that. SCHS was in its fifth year of program improvement for an additional four years. This meant that the school was labeled as low performing for nearly a decade while little was done schoolwide to address this fact.

The Freshman Preparatory Academy

Perhaps the most visible response I saw to stagnant academic performance came in 2010 (a year prior to my own study). In light of staff concerns about ninth graders who were at risk of dropping out, the school established a Freshman Preparatory Academy (FPA). The FPA was designed collaboratively by the core teachers for each calendar track, and each cohort of teachers established key principles they felt should guide their instruction with the new high school students. Although common pacing plans were established in core content areas such as math and English, the principal at the time also stressed teacher creativity. He encouraged teachers to pilot new pedagogical approaches that can be adapted by others if empirical data—as measured primarily by test scores—demonstrated student achievement. Although he did not speak directly to using phones and games in classrooms, I interpreted his mandate for creative change as encouragement to explore the goals of the FPA within my own research. Midway through the year, after successfully helping to launch the FPA, this principal (at the time, my seventh in so many years) officially resigned due to reasons unrelated to the focus of this study. The FPA persisted despite his absence, but a lack of leadership made teacher collaboration difficult.

Inconsistency and Movement on Campus

With the ebb and flow of administrative leadership at nearly every level from the federal government to the assistant principals who observed and evaluated my teaching, policies and their execution played an important but sometimes counterproductive role at SCHS. For example, during the years of this study, a new policy was enacted regarding where students could enter the campus in the morning. Students previously were able to use several entrances on all sides of the campus, but Mr. Panza, who was in charge of discipline and security, reduced the number of student entrances to two—one gate on the north side of the school and one gate on the south side of the school—to make the job of security on campus easier. However, students in my class noted that security guards were still posted at the former entrances to turn away students, and when I visited the two working entrances one morning, I found massive crowds. Students were skateboarding, being dropped off, trying to get through gates, and occasionally overflowing into the street, and (at least on my visits) they were supervised by only two adults.

A few years earlier, another major campus policy began to change the way I perceived space and interactions within it. The old hall pass was a sheet of paper affixed to a clipboard, but this was replaced by a new hall pass—a bright green and yellow vest like those worn by construction workers. Any students who had to leave classrooms to use the restroom or travel elsewhere on campus had to wear these new vests. The new vests would make it significantly easier for security guards to identify students who were approved to leave classrooms and students who might be considered trespassing.

As a researcher, I saw these new vests code the student population in a way that refocused adult attention on specific, visually apparent students. Many of my students felt that it was demeaning to have to wear a bright uniform just to use the bathroom. The vest was seen as another effort to dehumanize student experience at the school.

When the liquor store across the street from the campus began selling identical vests, a viable market in hall passes beyond teacher purview

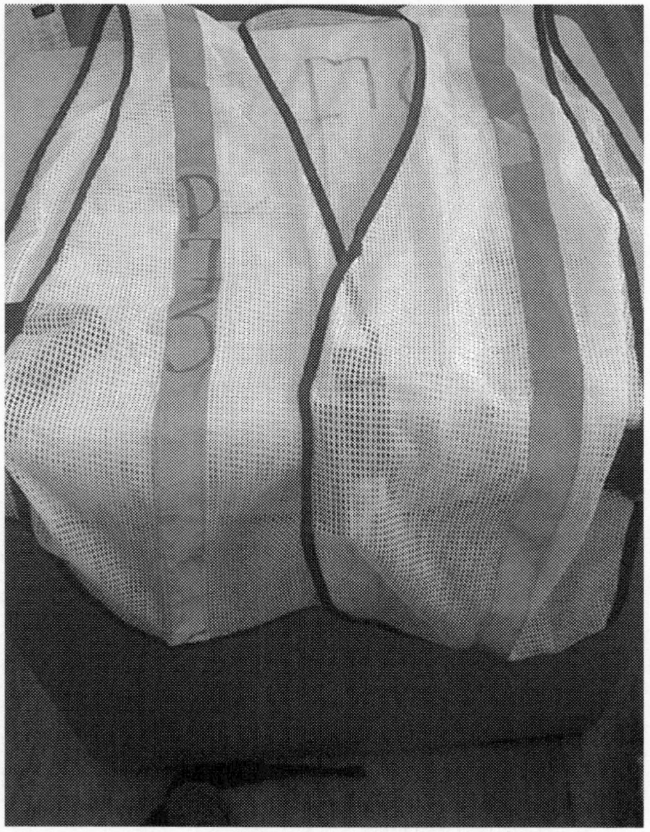

Figure 1.4
A green and yellow construction vest functioning as a school hall pass

was created, which undermined the discipline policies on campus. As a related side note, even though a school uniform was in place, students wore partial uniforms to school in order to get past the watchful eye of campus security.

It may not be the most professional of decisions I ever made, but I asked one of my artistic students to spray-paint a large black fist onto the back of my official classroom hall pass vest. Mr. Panza confiscated it and, even though it still functioned as a very bright hall pass, would not return it to me because a vest with an image on it would be "too confusing" for school security.

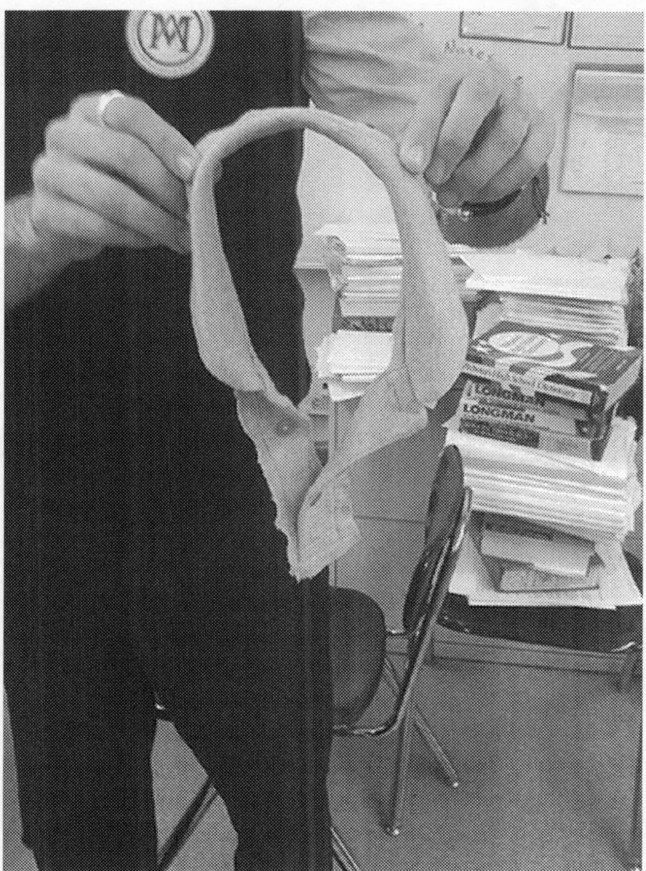

Figure 1.5
A student's workaround for the shirt collar required by the school's uniform policy

Figure 1.6
A customized hall pass

After my hall pass was confiscated, I sent students out of class with no pass and told them to explain to security guards (if stopped by them) that Mr. Garcia was waiting for his pass to be returned. After the weeks of being vestless in the halls, only one of my students was ever asked why he was out of class without a vest. In effect, the new passes were another inconsistently applied campus policy.

The Inconsistency of Time

Just as campus policies for leaving class often shifted, the measurement of time was also slippery. I saw SCHS's use of time denigrate students in ways that can be read as classist, racist, and apathetic toward the needs of urban youth. In an ironic reversal of the racist notion of "colored people's time," students at SCHS were constantly left uncertain about when a bell would ring and when their lunch would begin or end. Another microaggression left unchecked.

During the year of this study, the bell schedule changed three times in the first two months of the year. Why did the schedule change so often? Usually, it was caused by simple arithmetic errors that were left unnoticed over the summer. As a memo sent to teachers noted:

> Due to a mishap last spring with the district bell schedule design software, our current schedule is 18 minutes too short for our regular length days. Therefore, 2 minutes will be added to each passing period, and 1 minute will be added to each class period.

For the third bell change, a slightly *more* confusing notification was sent to teachers that required an approval vote from faculty regarding the new allocation of school minutes:

> Please find the NEW Bell Schedule attached. Please note that since staff voted to eliminate Advisory on Monday, Tuesday, and Friday, there are <u>6 minutes</u> added on to Period 4 on Monday and Wednesday and <u>2 minutes</u> added on to Period 4 on Tuesdays to allow for school announcements. On Wednesday and Thursday, announcements will continue to be at the end of Advisory after lunch. Please remember this new bell schedule will not start until a week from Monday on October 17.

"Advisory" was a counseling period where students could gain mentoring support from teachers, share their learning experiences, gain study skills, and prepare their postsecondary plans. During this year, teachers received no clear guidelines on how to use the advisory time, and with the above announcement, advisory was cut down to meeting only twice a week. For students needing mentorship and opportunities to connect with adult and peer mentors, this move nearly eliminated the advisory opportunity all together.

In addition to the structural changes, SCHS had a problem ringing bells at the right time. Starting on the first day of school, the bells rang at erratic times, anywhere from one to seven minutes before or after the "official" time. Minutes ebbed and flowed uncertainly as teachers and students eyed the clock, uncertain about whether class should end or continue. This felt like a professional slight to teachers who plan for consistency and depend on bells ringing correctly in order to end class in an orderly fashion.

These examples of disrupting consistent learning time affect how students perceive their schooling as valued and important. Like receiving truancy tickets for coming to school, the discrepancies between school policies and the lived realities of students and teachers can be striking. For the urban youth at SCHS, these inconsistencies tell students that adults do not care enough about them to run the school properly. They reinforce messages of shoddy schooling perpetuated in the media and stereotypes about urban communities. This is the way our schools define a new and even more racist understanding of "colored people's time."

Technology and Unspoken Policies in Campus Life

The following chapters of the book closely scrutinize the role played by technology at SCHS, but it is important to pause here to consider the policy implications of electronics at the school. There was no clear delineation of what was and was not allowed with regard to technology use during my years at the school. Although national policies and research trumpeted the potential of digital tools to transform and level the educational playing field, district and school policies seemed to be more concerned about preventing classroom disruptions, protecting Wi-Fi passwords from students, and fretting about teachers' use of social networks. They reflected a paradigm of technology that was archaic and campus policies that were unclear and largely undiscussed. The landscape of technology use in classrooms has shifted significantly in some ways since I left SCHS, but inconsistency reigns supreme when it comes to implementation and articulation of policies at the school and district levels.

At the same time that the school's assistant principals advocated incorporating technology within classrooms, school policy suggested that the electronic devices that students were most comfortable with were seen as distractions rather than useful learning tools. The policy constructed these tools as counterpedagogical, despite research illustrating the learning opportunities that phones and handheld devices can offer.[20]

The campus policy on technology not only limited student engagement in online activities but also provided barriers to students seeking out information on platforms that students are already literate in. Both the environment and the ways that information seekers feel within an environment affect the process of engagement and learning opportunities. The school's physical environment, resources, and policies help perpetuate the digital divide. When computers look dilapidated, are covered with graffiti, and load applications slowly, the message to students is that the school doesn't care about student learning opportunities. The civic lesson is one of inequality: as students in poor schools, they do not deserve the same opportunities as students in more affluent schools.

Likewise, even though digital technology was not allowed for most students, adults relied on electronics for control. Adult voices were amplified by PA systems and bullhorns, and adults wielded technology to enforce discipline in the school. I often saw adults use technology at SCHS to control the campus. In chapter 3, I discuss how giving iPods to my students disrupted classroom order (and led me—like the other adults discussed in this chapter—to undermine my own pedagogical best interests). Based on my observations of adults and school policies, rudimentary technology seems to get in the way of student voices. The larger school climate is affected by adult actions and inconsistency, and teachers must foster powerful relationships within our classrooms.

Graffiti and Meeting the Most Present Student I Never Met

As this tour of South Central High School comes to an end, we find ourselves in the bungalow where my first class met. Although not directly addressing campus policies, in this section I return to the students and their lasting presence within the school. During my first year of teaching, a spectral presence named Fonce Woddy invaded my classroom. When I first entered the classroom, it looked a bit worn from years of use but fairly clean, except that the words "Fonce Woddy" were written—on many desks, on the inside of the room's doorjamb, inside a drawer in the teacher's desk, on a trashcan, in the classroom circuit breakers, and on a computer monitor. In my first week, I pulled down the projector screen mounted in front of the classroom to project a video for the class and discovered that a huge Fonce Woddy tag was scrawled across the screen.

Numerous studies have examined tagging graffiti, the role it plays in identity and culture, and both the negative and positive aspects of this largely illegal form of art.[21] Rival gangs tagged the walls of local businesses, and students and nationally known artists contributed to on-campus graffiti murals. The presence of graffiti as an art form, a mode of communication, and a cultural label could be seen nearly everywhere on and off campus in our neighborhood. Perhaps graffiti added to outsider perceptions of the school as "urban," but it was a clear part of student culture.

Figure 1.7
Fonce Woddy's tag on my trashcan

Students read walls as updates on current rivalries, appreciated complex lettering and painting, and were annoyed when a textbook or class novel was covered in tags. Graffiti was always part of classroom life.

Conclusion: Looking beyond Mismatches

Energy. On any given day, the space of South Central High School swirls with bodies and excitement. The sight of so many students ready and prepared for challenges in classrooms can be beautiful and rejuvenating. Given the density of students at the school, SCHS can't help but be busy at any

Figure 1.8
Fonce Woddy's tag on the classroom's circuit breaker box

given moment. And it's not just the students who exude energy and enthusiasm. Although teaching at SCHS can be hard, there is a reason I spent my entire secondary teaching career at the school. My work as a part of the teaching labor force—as an educator, a learner, and an organizer—was valued by my peers and by the local community. It was fun to teach at SCHS, and it was rewarding. The politics of being a teacher, of being uncertain about job layoffs each year, of sorting out shifting requirements within my classroom: that stuff wasn't fun, and I strongly believe that it is one of the biggest ways we push quality teachers out of the profession on an annual basis.[22]

As frustrating as some of the experiences of being a teacher or a student at SCHS can be, the school thrives on possibilities. Looking at policy inconsistencies, daily microaggressions, and gaps in technology might paint a picture of a school that operates shakily on a day-to-day basis. It gets better, and understanding how the school functions allows us to dive more fully into the social lives of youth and their wireless practices in the next chapter and then my own classroom's foray into mobile-mediated learning in the remainder of the book.

Throughout this chapter, I've offered glimpses of what life looks like at SCHS through demographic and "achievement" data about the school, the policies that govern what takes place in classrooms and around the campus, and anecdotes about student experiences. And yet these textual descriptions can never capture the essence of what makes SCHS tick—the vibrant student life, the hunt for a parking space in the school's parking lot when you're running late, the adolescent-like joy teachers experience when driving the custodian's cart across campus to haul heavy materials from one room to another.

SCHS and other urban schools around the country have a lot of work to do to improve. And yet even this chapter's surface-level look at SCHS highlights how some student intentions are thwarted by adult actions. Ultimately, the mismatch between what students *need* as developing members of society and what adults *provide* within the school is one of the key tensions that emerged from my research. Looking at how students understand the campus space (as discussed earlier), classroom time (discussed in chapter 2), and uses of in-class technology (chapters 3 and 4) illustrates that a school may appear to be doing its best for its students but consistently failing them. Although a school can send a strong message about adult authority with a highly visible police presence (such as the line of police vehicles that greeted me during my first visit to SCHS), we can learn a lot about how to transform schools if we better understand what youth are doing and build trust in our day-to-day interactions.

2 The Always-On Student: Why Socializing Can't Wait until Passing Period

Antero: Can you tell me how much time you spend at school on your phone?

Minerva: The whole day!

Antero: You spend the *whole* day on the phone?

Minerva: Mostly.

Antero: What do you do?

Minerva: Everything.

Although I interviewed more than one hundred students to gather the data that informs this book, one of my ninth-grade students, Minerva, got to the heart of mobile media use in schools today. For students, nearly all decisions at all hours of the school day—deciding where to meet during periods, coordinating bathroom breaks, staying connected with peers at other schools or on other tracks of the school's calendar, and staying abreast of various pop culture updates—are channeled through their phones.

In many ways, the pervasiveness of mobile use in schools isn't surprising: it mirrors how many adults spend their time with phones. As I write at my desk or participate in a meeting, my mobile device is next to me or in a pocket, occasionally vibrating with a query from a friend, a reminder from a family member, or some other notification of the world beyond the immediately present one. Media scholar Douglas Rushkoff warns of the "present shock" of the constant background noise of updates that funnel into the lives we live in the moment.[1] The *amount* of time that kids spend on their phones during school hours likely isn't very different from adult usage. According to data from the Pew Internet and American Life Project, mobile devices are used frequently in both home and workplace contexts by people of all ages.[2] And although this data emphasizes that teens are "much more likely to use their phones to avoid boredom ... and also to avoid other people,"[3] the findings show that teenage socialization practices with mobile devices do not make them a demographic that is necessarily more connected than working adults.

As I began asking questions of the students who participated in focus groups for my study, they quickly corroborated the assumption teachers made that most students at South Central High School were utilizing mobile media devices. In all of the focus groups that I conducted, only eight students claimed not to have a mobile device, and even these students noted that they often used other students' devices to contact friends and family when necessary. All other participants owned a mobile device. Statistics on tech use and ownership corroborated these student claims. Surveys of teens from around the same time that I conducted this study found that more than three quarters of American teens owned smartphones and 79 percent of teens claimed to own iPods or mp3 players.[4] In the few years since completing this research, these numbers have increased. In addition, recent research has illuminated a growing "smartphone-dependent" population that can access the Internet from home only via their mobile devices.[5] Although only 7 percent of the U.S. population, this group is made up of young adults, nonwhites, and individuals with low household incomes and levels of education. The students at SCHS were likely to be smartphone dependent, which reinforced the existence of a "participation gap" between youth who can regularly access online activities through laptops and computer and those who rely on small screens and keyboards.[6] (A detailed description of my methodology for these focus groups and the findings throughout this chapter can be found in appendix A.)

When discussing student digital life at South Central High School with students, I learned that during some parts of their daily schedules they did not use their phones. However, as is discussed below, even these instances were carefully planned through prior mobile media use.

Although I focus much of this chapter on mobile media use specifically at South Central High School, I believe this data can be used to explore mobile media during the hours of 8 a.m. to 3 p.m. in schools across the country.

Before diving more fully into my conversations with SCHS students, I should confess that I conducted the interviews and observations for this chapter with preconceived feelings about the role that mobile devices play within schools. I provide more details about my researcher positionality in this book's appendices, but as a classroom teacher, I initially felt that these devices often interfered with my curriculum and instruction during class time. My goal with the data presented here is to attempt to see mobile use from a student perspective and to help other adults also understand not only *what* students are using their devices for but *why* and how these devices change youth culture today.

Proximity and Purposes of Mobile-Device Use: A Conflict in Policy and Space

Of all of the findings this research revealed, most surprising is the way that youth mobile use runs almost completely counter to adult provisional use of mobile devices. Students want to use mobile devices in ways that are opposite to how and when they are allowed to. At South Central High School, students generally are allowed to use mobile devices during lunch and nutrition periods during the day and generally are not allowed to use their mobile devices during class time.[7] Pretty simple and intuitive, right? "During class time, put away your phones. You can use them at lunch."

This type of structure may be problematic if teachers intend to use mobile devices as part of the learning process, which has its own pitfalls (as I show in chapters 3 and 4). The use-your-phones-on-your-own-time model is most problematic in how it illustrates that adults largely misunderstand youth engagement and mobile media use.

In conducting focus groups with students, observing campus spaces, and sitting in various classrooms, I began to see that there was a tension between what transpires during class periods and what happens between class periods. Most students didn't want to use mobile devices during lunchtime or nutrition period, and some students who received a text message or call during lunch waited until they were heading back to their classes to respond. So the time that students were allowed to use their phones was precisely the time they did not want to use their phones. This might seem counterintuitive to adults, but it makes sense when looking at what students do on their phones when they pretend to read a book or search for something in their backpack.

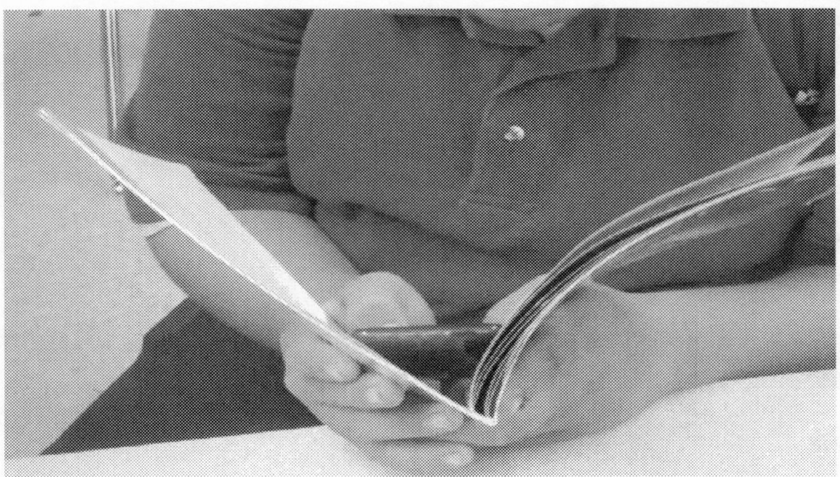

Figure 2.1
A student demonstrating how to use a phone while pretending to read a book

In classes, students used their mobile devices to socialize with each other, update each other on the latest *chisme* (gossip), and discuss meeting places and upcoming activities. After doing all of this organizing, planning, and chatting, teens didn't want to waste the precious minutes of their lunch break in isolated relationships with their devices. They used this extracurricular time to talk with friends and socialize without the mediation of handheld devices. My ninth-grade students told me the purposes behind *when* students use mobile devices:

Antero: Do you use it as much during breaks?

Dede: No.

Minerva: Uh-uh.

Antero: Why not, Dede?

Dede: Because you're interested in something else.

Minerva: Yeah.

Dede: Now that people are out there, you can put it away.

Solomon: Yeah, you don't need the phone to go out and start talking.

Minerva: It's just that when we're in class time, we got nothing else to do.

Antero: You've got *nothing* else to do? [laughs]

Solomon: You text someone instead of walking up to someone during class time.

Antero: Help me understand. Teachers think you should have your phones put away during class and take them out during lunch and nutrition. Right?

Minerva: Right.

Antero: But students use their phones during class and mostly put them away during lunch. Is that right?

Minerva: Cuz at lunch you're having fun, and people are around.

Ras: There's so much to do, like, during lunch that I might say, "Hold up. Imma check this message during fourth period."

How we organize schools today limits how much time students get to socialize with one another. Because educators want students to focus on academics during class time, socialization is pushed to the margins of the school day—passing periods, lunch time, before and after school. However, students see time as much more fluid than the binaries of adults. Instead of academics and socializing being two separate areas of focus, they intermingle these two times. This fluidity is reflected in many college-educated adults' workplaces. However, the lessons that adults could teach students about how to negotiate time are not only ignored but actively blocked by current school structures. The contradiction between how teens use their time and how adults reinforce the rules of time leads students to use their mobile devices to plan their precious moments of face-to-face socializing.

The comments like the ones from my students above were unanimously echoed by students in all of my focus groups, confirming that school time at South Central High School is interpreted differently by students and adults. Considering the contested space between student practice (using mobile devices for socializing during class time) and adult policy (reserving mobile devices for use during lunch and nutrition breaks) from a student perspective, it seems that students negotiate social time in ways that mimic adult interactions. I felt conflicted as a researcher and teacher when I saw students using devices at times not allowed by school policy. However, when I am in a meeting, I discreetly send text messages, respond to requests, and quickly scan my text messages, e-mails, and social networks for information immediately pertinent to me (and I have received negative comments about this behavior). Seeing my own activities mirrored and reflected back to me by students challenged my teacher perspective.

Although school policies are in place to hold students' attention during class time, adults need to consider the lessons we are delivering with these policies. I can think of occasional times at dinners, happy hours, and other social outings where a friend or family member's phone use takes him or her *out* of the present moment in ways that many of us would consider rude. Sherry Turkle's research has looked at how the alienating role of technology continues to affect generations of young people and their relationships with each other.[8] People of all ages are still figuring out how to use their devices in social contexts. And as mobile devices move to our wrists and to various devices through the silent takeover of the "Internet of Things,"[9] the constant prodding for our attention continues.

By asking students to interact with their mobile devices only during social times, we are focusing on the technological affordances of complex, online devices and are avoiding a shared examination of how these devices affect relationships. For educators, disciplinary issues arise, and we can spend time discussing whether student texting in class signals a problem related to an unengaging curriculum. However, schools that want to be up to speed on a quickly shifting world where students socialize with peers must contend with the fact that students are breaking school policies that interfere with student interpretations of school time.

Similarly, students at SCHS uphold an implied social norm about the amount of time that can be taken to respond to a text message. Most students I talked to said that they try to respond to text messages as quickly as possible, even if they are in the middle of a class. One student noted, "At least you should say you're busy and can't text right now. Otherwise, that's rude." Immediate responses to in-class texts are so accepted, in fact, that one student said that if someone doesn't respond right away, "It probably means their phone was taken away." Several students nodded in agreement. This cultural norm can frustrate students who feel that the demands on their attention are not manageable. Most students discussing this expectation of immediate response found it to be exhausting. Further, these students were aware of rules about no phone use in class but felt that the rules governing their social lives were more meaningful than the school's oppressive policies. Evidently, all of the rules regarding student behavior were not adult-instituted; youth-created policies on mobile use also were demanding.

Asking my own students why they use their phones when they know they are not supposed to, Solomon shrugged and said, "It's fun."

Ras: Just because.

Minerva: Because if it's vibrating, you need to see who it is.

Antero: Why can't you wait?

Minerva: It could be important.

By silently challenging the rules and expectations of adult-dictated policies and following their own rules, today's students exert their agency through discrete mobile use. The rhetoric of frustration from teachers is clear: teachers I spoke with echoed popular media sentiments that students are "distracted," "rebellious," or "rude." But these comments came from a challenged authority commenting on school rules. Because of the ubiquity and compactness of mobile media, students are redefining school behavior. Their social rules are superseding school policy.

Similarly, several students offered disconfirming evidence of being annoyed when friends texted them too often in class. Minerva, for instance, said that she tried to respond to friends within five minutes, but she also said she is frustrated when, "They [friends], like, send you one text, and if you don't write back in like two minutes, they send you, like, three fucking messages. That's dumb." Discussing the reasons behind her frustration, Minerva explained that sometimes, "I just want to pay attention to what's going on [in class]." Minerva's experiences highlight the student perspective that time in school is fluid: it is both academic and social, and the boundaries of these two uses of time in school are not dictated by school bells or peer demands.

Not a Phone

Continuing the analysis of how students use their mobile devices, it is necessary to reflect on the vocabulary we use for the things we carry in our pockets and purses. How often are the "phones" we carry in our pockets actually used in the "old school" ways that telephones used to function? As a colleague noted, people in many other countries use the word *mobile* for what Americans call *cell phone*.

Considering how infrequently we actually hold the devices to our ears as receivers, calling mobile devices "phones" seems like a misnomer. Recognizing this, I began my work by asking students what kinds of things are "mobile." The inner-city youth I worked with identified phones, iPods, and game devices like Nintendo's DS as typical "mobile" devices that might be used numerous times throughout the day. Two students I interviewed suggested iPads and tablets (two years before the Los Angeles Unified School District purchased iPads for every student in the district). Not one student identified a laptop computer as a mobile device.

Sound, Talk, Privacy

For a campus that is always connected, SCHS is a generally quiet one when students are in their classes (aside from the occasional helicopter overhead). The most frequent sound heard on the mobile-enabled campus is the warble of music from earphones that are hung over ears—but with the speakers positioned outward to project publicly the music students experience. Rather than using headphones to curate personal listening experiences, students blast their music at maximum volume, using earbuds as miniature speakers to DJ for nearby peers. Similarly, students roaming the halls often played a customized ringtone to echo and be heard by students in classrooms with open doors.

Despite the prevalence of public curation through mobile sounds, students had very different visions of how private information should be shared. One common concern among students was whether information could be overheard or seen. For example, in one focus group, a student said that when he was talking with "his girl," he did not want his classmates to hear their entire exchange, even if snippets of his end of the conversation were audible. As such, both texts and phone calls in schools can be challenging for students. Phone calls can be overheard, and the emotions of the speakers inferred. As one student noted, "With a phone call, you don't want nobody to see it." Likewise, text messages invite wayward eyes to peek at the exchanges. Both auditory and visual communications in the school are found at the divide between private and public.

For many of my students, text messages and Facebook messages that they assumed were private have been forwarded to third parties and beyond in ways that have been hurtful and embarrassing. Dante and Katherine recounted an incident when a classmate's nude photograph was forwarded to many students at the school and perhaps beyond. These troublesome instances of teen "sexting" are among the many student communication practices that are shaped by mobile devices.[10] In fact, the ethics of mobile devices and students' understanding of privacy have grown increasingly complex, as researcher Carrie James explores in *Disconnected: Youth, New Media, and the Ethics Gap*.[11] Because it is relatively easy to forward content from one device to another, students relate incidents like the one described above as a warning about some forms of mobile use. In this sense, many students balance public and private conversations through mobile devices. For conversations of the most sensitive and immediate nature (those that students feel can't wait to be shared in person), many students quietly whisper phone calls in corners and bathrooms. However, the content of a conversation or an informal exchange via text or Facebook often is edited to ensure it appears innocuous enough to be shared with a larger, unknown audience. Many students profess love and also write cryptically in public Facebook posts that both encourage dialogue and occlude explicit meaning from all but a handful of close confidants.

In the introduction to a book for teachers in the digital age,[12] I shared my thoughts on seeing many of my high school students migrate from MySpace to Facebook (a divide that initially was tied to socioeconomic and racial demographics).[13] Students used the comments section of a Facebook post as the equivalent of an instant messaging or chat client. It wasn't uncommon for a student's post to generate forty to fifty different comments within a short amount of time—an entire archived conversation. Initially, I saw this use of Facebook as wrong. *That's just not how Facebook is supposed to be used*, I thought. But eventually, I came to understand that the "networked privacy" of students revealed a new paradigm for using digital tools. The contexts of digital tools and the ways they are used depend on who utilizes them: what looked wrong to me was simply a different kind of practice for my students.[14]

School Tasks, School Time. Social Tasks, Social Time

In talking with students and observing how they relied on mobile devices for social purposes at school, it became clear to me that their use of mobile devices in school was not simply about student pushback on existing policy. Instead, the ways that they engaged in mobile-mediated social practices during school hours was redefining how they used time, how they understood time, and how social time and school time were elastic and overlapping. Students are not sending text messages *instead* of participating in academic work. Instead, they see these two areas as overlapping and as occurring concurrently. As a researcher focused on the literacies of young people, I am interested in how students engage in literacy practices in two parallel spaces throughout the day. Numerous studies have looked at the academic affordances of texting, but I would add to this corpus the notion that *time* in schools oscillates haphazardly between academic and social. Further, when students are communicating with friends and looking at social networks, they are engaged in robust literacy practices.

In recognizing how student socializing interferes with time that is otherwise primarily academic and defined by adults, it is worth considering how students try to balance their roles as social agents with peers and as students within adult-run classrooms. Talking with my students days before the school's annual state standardized test, I learned that many students anticipated using the testing days as an opportunity to catch up with friends through mobile devices. Many students said that even when answering test questions, they also would listen to music. In general, there are conflicting visions about time and its use within school spaces. Mobile media is seen as a resource for shifting the nature of time in academic spaces to function both socially and academically. It is a duality that reinforces the multitasking behaviors that stereotype or vilify the learning habits of today's youth.[15]

Although such seemingly constant use of mobile devices echoes Minerva's comments that begin this chapter, my observations show focused pockets of engagement with both phones and instructional content at SCHS. Most students said they were willing to respond to text messages or check updates from their social networks with several minutes of delays. Essentially, when students felt bored or noticed a lull within the classroom structure (regardless of teachers' lesson plans), they shifted their attention back to their mobile devices. Although some students stated that they needed to respond to peers' text messages immediately, "they obviously know you can't get back" right away, as one student put it. This spectrum of responses illustrates that although student perspectives tend to coalesce around specific practices, their mobile ways of being at SCHS are not monolithic.

On Using Mobile Devices Academically

Starting in the next chapter, the scope of my research shifts from understanding the larger social landscape of the school and student relationships with their phones to a narrower focus on the needs of students in my own classroom. However, before getting to this conversation, I asked students for their thoughts about formally incorporating mobile devices in the classroom. Although there is a growing body of published strategies for using mobile devices in classrooms, these articles tend to discuss the pragmatics of using devices and apps while disregarding the orientations and expertise of young people. My focus groups with students showed that students have their own beliefs about possible uses for mobile devices for academic purposes in classrooms.

As with many other aspects of school designs and operations, adults rarely ask the key participants in schools—students—what they feel would benefit them. The expertise of youth is far too frequently overlooked or disregarded. For instance, discussing the learning implications of mobile devices, Minerva said they should be allowed in classrooms: "It's one way you can concentrate more at work, like listening to music. *I can*. And if they let you like text or something and let you do something that you want to do, then you could probably do something that they want you to do. You get me?"

In one discussion with students about the ways that teachers could use mobile devices for academic purposes, I recorded the following exchange with my ninth graders:

Antero: Can you think of a time you would text an adult?

Dante: I can.

Antero: Go ahead.

Dante: Like if you were absent and want to catch up.

Minerva: We ain't gonna do that!

Antero: If everybody had a phone and I said, "Text me your answer now," would you do that?

Minerva: Yeah, that would be tight. Then we would get our points and shit.

Dante: I wouldn't even give you my number.

Antero: Dante, you bring up a really good point. Are you worried if I had your phone number, I would use it in a way that would make you uncomfortable?

Dante: Yeah.

Antero: So you'd feel weird if I texted you?

Dante: I guess. I mean, you're my teacher. I don't know you like that.

Antero: So is a text message more personal than a phone call?

Dante: No, I'd rather talk on the phone to someone but rather *you* text me than call.

Minerva: I feel weird and uncomfortable and awkward talking on the phone to you, so I'd rather you just text me.

Antero: Okay. Well, have you ever been on Facebook with a teacher?

Dede: Yours.

Antero: Oh, yeah. [laughs]

Ras: Really?

Antero: Yep. Remember, you can be my Facebook friend. The info is on our syllabus.

Dante: Naw. You gonna burn me out.

Antero: What do you mean?

Dante: You'd find out too much about me.

Minerva: Yeah.

The nuanced nature of participatory media tools like phones and social networks means that there is no clear endorsement of digital device use in classes by students. Minerva was excited that a phone means "we would get our points," but she also was wary like Dante that a teacher would "find out too much" in a social network like Facebook. Although some students saw opportunities to engage academically if a teacher opened a space for mobile use as part of the classroom practice, this exchange highlights the fact that student opinions vary. Their comments were tied to specific components of my own class structure, but they exemplify the differing opinions I encountered in students at various schools. Some students, like Minerva, saw using mobile devices in classes as opportunities to increase their class participation, but many others, like Dante, were concerned about how teacher access to their mobile devices could uncomfortably extend the student-teacher relationship. Finally, even the various applications available on most phones are seen differently. Minerva seemed initially eager to use her phone to text responses to the class, but she would "feel weird and uncomfortable and awkward" speaking with a teacher via a phone call.

Because mobile devices were strictly the domain of social circles at SCHS, using them in an academic context felt alien and threatening for many students. But the fact that these devices connect students socially while in classrooms is an important feature that could be used to rethink student orientation to learning in schools. However, the skepticism, distrust, and apprehension that students expressed revealed that simply placing mobile devices within a classroom would do nothing to build stronger relationships or academic outcomes. Data from this study directly influenced how I approach classroom design with mobile devices in the next chapter.

In exploring how mobile devices affect classroom learning, I asked students if they were already communicating or socializing with teachers through digital devices. Some students were in social networks and texted teachers that they trusted. "She's just different from other teachers, way cooler," one student said about the teacher she felt comfortable texting. Aside from these few teachers, most students said that they would feel uncomfortable communicating with teachers with mobile devices. Although students saw mobile devices as tools for communication with peers and family members, the devices did not signal the same opportunities for communicating with educators. In fact, this aspect of student use reinforces traditional adult-student interactions. The introduction of mobile-device use on campus technically introduced a new means of communication between teachers and students, but in general this opportunity was ignored. Students did not see their devices as connecting to their classrooms. Although student time mixed fluidly between academic time and social time, the tool that facilitated this fluctuation was used solely for social purposes.

Finally, in exploring trust and the academic opportunities for mobile devices, the social and academic divide in schools must be addressed. In the focus groups for this study, I tried to incorporate representatives from as many diverse groups of SCHS students as possible. However, as a school that is comprised solely of black and Latino students, SCHS can show the significant ways that mobile-device penetration affects student social life. Research identifies these students as richly literate during their "out-of-school time,"[16] even though the same students perform poorly in academic contexts. If teachers can bridge social time and academic time within schools and utilize students' vast mobile literacy skills, perhaps they can help students increase their academic engagement and improve academic outcomes. My own attempts at this are detailed in the following chapters.

Resources and Finances

Mobile devices are expensive. I focused parts of my research on exploring how students were able to afford mobile devices and texting plans and what strategies they used to save on these costs. Because students at SCHS were and continue to be seen as "economically disadvantaged"[17] and because many of my students needed to work to help support their families, the financial challenges of owning and using mobile devices were particularly worth looking at. Most students told me that they bought used phones or received new phones by signing multiyear contracts with cell phone carriers. These two different avenues to smartphone ownership highlighted two different financial strategies. On the one hand, many students who had multiyear phone contracts said that they got minutes for their devices on a "family plan." On the other hand, other students paid lower rates through popular month-to-month carriers.[18] Ownership of these devices is required for engaging socially in schools, but it is intriguing to think of how this ownership extends beyond the social bonds of the school. According to focus groups, when students have to share monthly minutes with family members, parental decisions and budgetary constraints often affected their mobile use.

Students also took advantage, as much as possible, of applications that can replace carrier fees for texting and calling. Additionally, students' awareness of the possibilities for ownership and the limitations of plans points to their ability to negotiate cost and plan limitations. Solomon, for example, showed me several applications that he uses to text for free. He said he has had less success with the apps that allow for free calling than the texting apps but comfortably changed from one texting app to another, demonstrating by sending random missives to friends. In heeding the lesson that Solomon provided, I later used the free texting app with my ninth graders to communicate using iPods (chapter 3).

Youth Knowledge

Each time I buy a new mobile device (every few years, when the devices seem to break just as my mobile carrier's contract is up for renewal), I find myself flummoxed by the functional skills each device requires. I asked the students in this study if they, too, ever had difficulty figuring out or using their devices:

Student 1: No. Only adults.

Student 2: Yeah, my moms be having trouble, but now she's got the hang out it.

Student 1: Every phone's got the same shit. It's just in a little different ways that takes a bit of time to figure out.

Because students often look at each other's screens, sometimes share devices to listen to music, and often upgrade to newer devices when financially available, youth technological knowledge is implicit and abundant. In two different focus groups, students shared ways they used features on their phone in rapid exchanges that, for me, were difficult to follow. The technical aspects of using these devices and talking about their technical aspects were such common knowledge that it was easily shared among peers and demonstrated. In most focus groups, I asked students to show me how they text messaged quickly in class. They demonstrated on several different devices, and I appreciated the sophisticated thumb tapping, finger swiping, and other hand gestures that are second nature to students. Watching their high-speed dexterity, I saw how these students' hands understand, interpret, and send out information in artistic bursts and seemingly haphazard taps. The physicality of this writing is significantly different from typing with all ten fingers on a keyboard or using a select set of digits when wielding a pen or pencil. The thumb-based writing encouraged in a digital age shifts the reflexes and muscular process of writing.

Similarly, students said that the popularity of certain devices nearly always was equated with price. The features that are prominently advertised are valued on these devices. Although they may not be the most practical features, most students wanted their phones to have "the 4g," a touchscreen, and GPS (Global Positioning System). But when one student was asked if she used the GPS on her phone, she said, "Hell, no. What I need to be using that for? You know we ain't going anywhere but the hood."

Aside from the phone's features, the other important factor in how students decided which devices to use was the cost of data and talk time based on various plans. Students seemed generally split. About half of the students had month-to-month plans that typically were paid for by a parent. The other half were part of a family plan in which data and calling minutes were shared in a pool with parents and siblings. Although both types of plans had limitations, students generally had strategies for the days near the end of the month when data or minutes ran low:

Antero: Do you ever run out?

Student: Yes.

Antero: So what do you do when I know you're about to run out?

Student: I stop using it as much.

Antero: You stop texting?

Student: Yeah.

Antero: So if someone texts you, you tell them you can't text right now?

Student: I tell them to, like, call me on my house phone.

Further, the locations that students found themselves in throughout the school day also affected student mobile use. Students told me that they deliberately went to bathrooms at various locations on campus to have more time to use their devices or to meet up with friends in ways that were preplanned through mobile media use. Additionally, students were keenly aware of where phone reception was weakest at the school. When asked if there were particular areas at school that had bad phone reception, students gave the same answers. One student said the school's basement, where the Junior Reserve Officer Training Corps (JROTC) program was housed, was known to have bad reception. Another student replied, "Yeah, in the JROTC building, you got to be in the one corner at the end to get reception." Students knew about this known space of reception in the basement and told me, "There's, like, half the class standing there." In following up, the students said that the JROTC chiefs were oblivious to the reason that students converged in one corner of the basement.

As noted before, student social interactions superseded adult academic expectations and policies. Students navigated the school space searching for mobile signals that could leverage and maximize their social engagement, with or without adult permission

Mobilizing Activism: Communicating a "for Real Thing"

As a portal for civic engagement, a mobile device can function as a personal system for organization and participation. During the time frame of my study, many of my students participated in a walkout and two campus sit-ins related to school budget cuts in California. Although teachers and administrators were aware that students communicated protest plans via mobile devices, the kinds of preparation involved in such practices have not been clearly articulated in prior research. During my years at SCHS, I witnessed numerous student walkouts at the school, and students leveraged text messaging features on mobile devices to plan and participate in these school-based civic protests. In this way, handheld technologies were integrated into civic identity in urban schools. After discussing in-class use of devices in my focus groups, I asked students how they became aware of the on-campus sit-ins and walkout activities that were organized by students:

Student 1: I got a text.

Antero: Who did you get a text from?

Student 1: I don't know. Just random people forward it to you.

Student 2: It says, like, "Send it to your friends and their friends and their friends."

Student 3: Almost everyone knew and knew the day before.

Student 2: I got a text message and saw it on Facebook.

Antero: What did the message say?

Student 2: It said, "Walkout tomorrow after nutrition."

Antero: So how did you know what to do? Did you just follow the text's directions when it was the end of nutrition?

Student 1: People started walking to the yard.

Antero: So you could see them.

Student 1: Yeah, they were walking, so we knew it was a for real thing.

After students in this group saw other students acting, it was clear that this message detailed "a for real thing." There is a sense of chance and discretion for students when receiving these text messages that invite them into civic actions—activities that respond to and focus on issues related to their school life. Additionally, the time between receiving a text or Facebook message and the walkout was only one or two days, but no students expressed frustration at this short time frame. These students paint a picture of last-minute and sometimes faulty plans when using mobile devices for civic engagement. Their hesitancy to assume that the texts could be trusted signaled ways that students felt disconnected from the sources of information and needed more trust before they would act.

What is also striking about the way that mobile media informs and guides student organizing is that it does so anonymously. I talked to several students who authored the initial text messages announcing school walkouts that were circulated among peers anonymously. However, most students did not know or ask about the text's origins. In this way, text messages about walkouts and sit-ins seem to run counter to many of the general tenets of how students understand and utilize their devices at schools—that mobile devices are extremely personal. With texts related to student-organized activism, the information is depersonalized:

Antero: I've heard students tell me it's going to happen again tomorrow?

Student 1: Yeah.

Antero: How do you know that?

Student 1: Cuz it went around again.

Antero: You already got it?

Student 1: Yeah, I got it back on Wednesday.

Antero: But you don't know who sent it?

Student 1: No, it's like this little survey thing you're doing now. You said it's anony- ...

Student 2: Anonymous.

Student 1: We don't know who started it.

Student 2: We just got it.

The anonymity of these messages can be seen as problematic when trying to tease out the actual civic engagement experienced by students. For most students participating in the school walkout and sit-ins, information about these events was communicated through anonymous text messages. Although marginal participation in activism and grassroots efforts are key ways to add to public engagement and important steps for youth to see avenues for more engaged, future civic participation,[19] the anonymity of these text messages ruptures student connection to identifiable groups or individuals. Instead, students said they felt united in these civic actions by seeing their peers participating in a "for real thing." The anonymous text message as tool for leveraging mass mobilization at school is a modern method of engagement and protest. Movements like Occupy and #Black-LivesMatter are notable examples of how digital publics leverage digital tools like Twitter and Facebook to organize, communicate updates, and provide live footage of protests and police responses. Similarly, online protests are often anonymous. Perhaps the most notorious such group is aptly called Anonymous, which staged physical-world protests against Scientology and virtual protests against the Stop Online Privacy Act while masking (both literally and figuratively) the identities of its participants.

After my focus group sessions, several students forwarded me the text message that spurred students to participate in a school sit-in. Here is one of the messages I received:

> FWD: FwD FWD FWD SIT DOWN THIS FRIDAY!!! AND HOPEFULLY ANOTHER WALK OUT!! FORWARD TO ALL STUDENTS DOESNT MATTER WHAT SCHOOL!!

This text message's defining characteristics speak to the nature of dialogue and communication in a digital age. The anonymity of the message (with the numerous forward notices) can be used to protect the text's original authors. The all capitalized letters that "shout" to the reader also signal that this text speaks from a place of urgency that differs from most student text messages and Facebook posts. Finally, it offers clear instructions for students, which encourages action. It is easy for students to enter the handful of keystrokes required to comply with the message's request to continue to forward the message to others.

Teachers Have Mobile Phones, Too

Although this chapter focuses on student use of mobile media, I want to discuss briefly how my students perceived adult mobile media use in school spaces. Students in one focus group said their math teacher used his cell phone just as much as the students and perceived adult mobile use as frequent and hypocritical. "It ain't fair that he's texting and I can't," one student told me. Student responses point to parallels between adult dispositions toward mobile media use and student behavior. As the student above noted, some teachers ignored the restrictions and rules they set for the students in their classroom and openly used their phones.

In some cases, teachers utilized mobile media to connect with the world that loomed beyond the walls of the classroom. Students discussed one teacher who brought her own Internet router to school to access blocked websites. Because this teacher "doesn't have a lock on it," students in her class and in rooms nearby used the teacher's router for their own Internet use, allowing students to access Facebook and music downloading programs without using their own data plans, which may cost them money or be limited to monthly quotas.

Trust and Deception with Adults

For students to engage with mobile media use responsibly in classes, they require teacher trust. If they do not feel that their teachers trust them, as reflected in classroom structure and pedagogy, their mobile use may not comply with teacher goals. Perhaps more than any other challenge that schools presently face with mobile devices, the issue of trust looms largest and does not have any simple or clear answers in policy changes. Asking students in focus groups if they felt comfortable about giving me their cell phone number, a student said, "We have to work on our trust." Although this echoes Dante's concern that I'd know "too much" about him if I were to see his Facebook page and Minerva's preference for texting than talking with a teacher, this student's comment also signals a pedagogical path forward. Much as this study explored the ways mobile devices were being used by students and could be leveraged in classes, this student reminded me that the relational aspects of learning trump other concerns. Unlike other pieces of technology, the persistence of "always on" mobile devices in classrooms means that a temporary ban of these devices positions students to act out against adult-created rules. Similarly, within many business settings that exist beyond students' school careers, young people will be required to understand appropriate and inappropriate times to engage in on-the-job mobile-device use.

In a different context, regarding student phone numbers, one student told me, "Sometimes you have to give your number to a teacher if they are going to call your parents." This statement led to the following:

Antero: Oh, and then they call you?

Student: Yeah.

Antero: And then what happens?

Student: I keep my phone on silent.

Antero: Do you answer trying to be your parent?

Student: No, but I have an older sister answer mine.

A phone number given under false pretenses is useless information. If the teacher does not think the number is the student's, he or she cannot use it for in-class purposes. If students do not trust teachers and teacher-imposed rules, the likelihood that mobile devices will be used for academic purposes diminishes. In addition to the above example, students shared other minor deceptive practices. Students in every focus group told me that they would often text "behind a book," "in my backpack," or under a desk. These in-the-moment deceptive practices were necessary for facilitating the fluid social-academic nature of student time in schools.

Conclusion: The Digital Twilight

Two months into the school year, on another hot August day, the South Central High School campus was locked down. Lockdowns—a word that was lifted directly from the prison system and reinforced feelings of control and incarceration on the campus—required all students and teachers to remain inside their classrooms until a security concern was safely dealt with. Because this was not an uncommon occurrence at SCHS, teachers and students knew what to do when a lockdown or "Code 1000" was announced over the school's speaker system. This particular August lockdown was unique because it was due to a bomb threat. And also because it lasted for five and a half hours.

After evacuating our classrooms and moving into the school's auditorium and gymnasium, students and adults passed the time chatting, watching television shows on their phones, calling friends, and playing games. Aside from the wasted school day, the crowd was calm and easy to manage, all things considered.

Four hours into the lockdown, however, the police notified all students, teachers, and administrators that phones needed to be temporarily turned off. They were about to broadcast a signal to attempt to disable any explosive devices, and the effort could potentially break or damage phones. Dutifully, everyone on campus pressed buttons, swiped screens, and toggled options to turn off their phones and gaming devices. The hour that followed is probably the longest stretch of time during school hours in the past decade that mobile-device use ceased. No other classroom management strategy, threat, or coercive tactic effectively convinced *every* student, teacher, and administrator to shut off their mobile devices as well as the power a harmful police signal did.

During the hour of digital twilight, the school did not enter some sort of analog bliss, students did not suddenly look at each other in new ways, and the school did not roll into anarchy. This brief window of disrupted mobile use at school did not seem to reveal that we could not function without technology, except for one thing—time. I wasn't aware that an hour had passed until I was able to turn on my phone again. Without a wristwatch and with a broken wall clock in the auditorium, I shifted into *feeling* time pass in the conversations with colleagues and the giggles I tried to elicit from distressed students around me. The majority of the students and teachers I interacted with told me they shared this same feeling.

Our relationships with mobile devices and "smart" watches extend far beyond socialization and beyond classroom life. Although the chapters that follow focus on how mobile media and games can leverage, promote, and incite civic action, the existing landscape of teen social use of mobile media in class guides students in school life in fundamental and overlooked ways.

Time, as students noted in my focus groups, moved fluidly from social to academic. Time also was delineated and noted by our mobile devices (particularly when the school's bell schedule could not be relied on). Phones mediated schooling and the ways that students experienced it. They also mediated the world beyond school. Just as academic preparedness is a clear objective of public schools today, so too is the need for students to see themselves as participants in the world beyond high school. Our school systems need to do a better job demonstrating and supporting social uses of phones. In shunning mobile devices in schools, we are creating an artificial experience that does not reflect how people produce, develop, and interact in the “real” world. Likewise, by encouraging students to use their mobile devices only during passing periods and lunch breaks, we are promoting behavior that runs counter to the norms of most groups of friends. I have been reminded more than once that checking an e-mail or replying to a text message at the dinner table is less than polite. These poor habits feel all too similar to the ways that students hide their devices behind books while pretending to read.

Mobile devices present unique opportunities for engagement, socialization, and learning, and students recognize this by using these devices for continually communicating with families and peers, planning, and accessing sites of trust. At the same time, the data collected during my research revealed that the presence of mobile devices in poorly managed classrooms sometimes resulted in chaos. I heard horror stories of mobile devices gone wild. Even in my own classroom, mobile-device use—by the persistent in-the-backpack texter and the student who answered calls during silent reading—was often frustrating.[20] Although the data I collected allowed me to take a close look at the campus through my students’ perspective, these devices played a lasting role in shaping how I grew as a teacher at SCHS. Even though phones have been present in classrooms for more than a decade, they are still relatively new disruptions to the archaic factory model of schooling. More than perhaps any other form of cultural wealth that students bring to school or perform within school spaces, mobile devices mediate interactions between peers, adults, and parents. Digital twilights are a utopic myth. Rather than forcing devices to be off and away, our schools need to understand how they have shifted culture and our students’ understanding of time.

3 Inside the Wireless English Classroom

Why *in* the Classroom?

On a warm Monday in May, I handed out an iPod Touch handheld device to each of the seventeen ninth graders in my third-period English class. Throughout the school year, we had experimented with student mobile devices in class, and in these final two months, I planned to focus on an in-class initiative with all students using the same devices in similar ways. Students showed the kind of excitement I'd hoped they would have for undertaking this experiment with me, and I left school that day feeling optimistic about the journey that we were embarking on. After I finished my work at South Central High School, I headed to the grocery store and was about to enter the store at 4:04 p.m., when I received a text message from Holden. His iPod had been stolen (but he had hidden his own phone, so it wasn't taken).

Holden's text was the first in a series of disruptive and troublesome events that changed my thinking about technology in the classroom. After less than a day, and my students were already down one iPod. Although many schools now provide students with highly durable protective cases for digital devices (perhaps distressing Apple designers with their disregard for aesthetics), such precautions cannot ensure that all devices are going to survive from year to year. Perhaps the biggest limitation on widespread implementation of mobile media in schools is the lived realities of many urban youth. The story of Holden's stolen iPod above was not extraordinary in terms of the challenges my students faced with mobile media within their campus and school community. A week later, Jay came to class visibly upset. As the bell rang and the students took their seats, Jay asked me if he

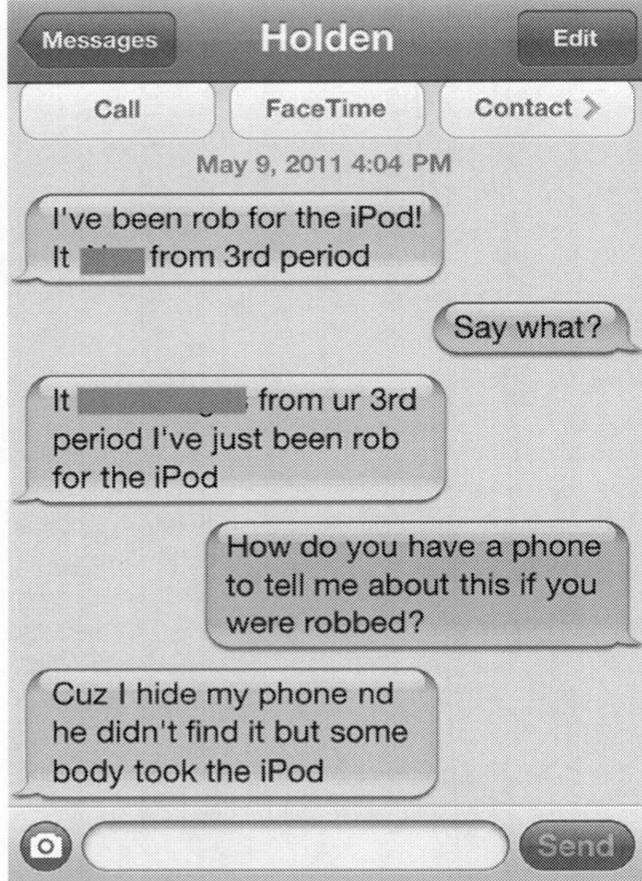

Figure 3.1
A text message from Holden explaining that his new school-provided iPod was stolen during third period

could talk to me outside. He explained that his locker was broken into during physical education, and his backpack and iPod were stolen.

Thefts were reported throughout the quarter. When I collected iPods from my seventeen students on the last day of school, seven were returned in functional condition, two were cracked, one displayed a frozen screen and was not usable, six were reported as stolen, and another one was missing (the phone numbers for a student who was withdrawn from school were not functioning). The low percentage of devices returned was worrisome for me as a teacher, particularly because the budget for this project was small. As a researcher, however, the data suggests the rich ways that mobile devices provided in classrooms may enter into the daily lives of campus and community members. Sure, kids can lie, and some students might have falsely claimed their devices were stolen, even while they continued to possess them. However, I took my students at their word, realizing that trust can go a long way for some students in a school space with a 60 percent dropout rate in any given year. Considering that only ten of the seventeen iPods were returned (and three of the ten were damaged), my attempts at integrating technology seamlessly into my classroom often fell short of my intended goals. I messed up frequently.

Why should I have bothered? As discussed in the previous chapter, SCHS was saturated with mobile devices, and youth socialization was augmented by the use of these devices (both in class and out). Much time was spent pressing, swiping, and posing with these devices, so it may not seem necessary to include phones as a deliberate component within the classroom. Further, these devices are not precious, even though they are costly and can break easily. When we spend time and energy fretting, fussing, and overprotecting mobile devices, we have less time to use them meaningfully.

If you bring up the topic of mobile devices in classrooms with any educator, you'll get a passionate response either validating or villainizing them. Mobile devices and participatory media are contentious topics within the field of education.[1] For some individuals, mobile devices are valid, necessary tools for equitable education, and for others, they distract students from academic learning. My experiences in my classroom moved across this spectrum and included unexpected challenges (such as thefts) and powerful moments of engagement. For every technical blunder on my part—such as attempting to lock students out of access to certain aspects of their devices or trying to lead the class through online research on the day the school's Internet was out (again)—there were promising moments of student growth. For all of the emphasis that my research placed on the mobile devices in my classroom, class members actually spent more time looking at each other than they had earlier in the year. Collaboration and in-person discussions were key parts of how the classroom changed. Today, I have even more reservations about how mobile media can positively impact English education and disrupt traditional learning practices.

Throughout this chapter, I detail specific ways that classroom life can be transformed by mobile devices. As schools globally are embracing one-to-one device programs within classrooms, teachers are expected to master the teaching challenges that come with these changes. This chapter is a slice of my personal story learning from and with my students in this new era of education.

Mobile devices might indeed be "Personal, Portable, and Pedestrian" (as Mizuko Ito and her colleagues suggest),[2] but they are significantly different from the traditional whole-class and small-group engagement that is typically found in classrooms today. I spend most of this chapter discussing the key participants that construct classroom experiences—students and teachers. Rather than describing what students *should* be doing in classrooms, I offer a look at what actually happened inside my own classroom. To do this, I first review what research with phones looks like when done in collaboration with students and detail what we did and why we did it. The rest of the chapter dives into the kinds of experiences that students had during our research projects. Although mobile phones were initially the focus of my inquiry, student voices became the most important part of the project. (I return to the role of adults in chapter 5, especially the struggles I had with parents and substitute teachers during the project.)

Reckoning with Research

My class needed to have a critical conversation—before I handed out the iPods and connected them to SCHS's Wi-Fi; before my students texted their homework to me, edited the school's *Wikipedia* page, and documented topics of inequity around the school space; and definitely before Holden's iPod was stolen on the same day that he received it.

Because I'd noticed how multidimensionally students used their phones—they listened to music on them, took photos of class notes, and texted me questions about class work—I was not interested in nonpervasive uses of mobile devices. Based on the data I shared in the previous chapter, I knew that mobile phones were valued by students and contributed in powerful ways to the school's culture. And so rather than focusing on the individual needs of students, I wanted to see what classwide implementation of a mobilized curriculum could look like.[3]

However, before I developed this curriculum and handed out these devices, my students had to understand that in this undertaking, we *all* were going to research the possibilities of these devices in schools. This was research we were doing together, and it followed the same content standards as any other unit I taught during the year. Further, I take my learning with students seriously. Their intelligence and capacity to educate and orient adults' thinking about learning in today's digital, participatory age were necessary to the success of this project. Although I've made mistakes in building trust, I have spent much of my time as a researcher doing this kind of work alongside young people.[4]

In preparing for the classroom inquiry on the use of mobile devices as tools for learning and engagement, I told students that our collective insights would help frame recommendations for what wireless devices can do for classroom practices. I was also honest about my own uncertainties about the results. Even though I had done some research, come up with some hypotheses, made some haphazard experiments with devices in classrooms, and conducted some interviews with students across campus (including most of the students in my classroom), I could not predict if our work would create more opportunities for distraction and disruption than for learning and innovation. Like a new teacher, I felt uncertain about whether I would be able to *control* the classroom. In hindsight, I realize that the issue of control—a central debate that swirls around the potential role of mobile devices in learning—reinforces and restricts the agency of young people. Eventually, my attempts at control were usurped, and similar issues are presently unfolding around the world in schools today.

As I handed out the seventeen iPod Touch devices, students excitedly started them up and I felt the buzz of possibility moving throughout the room. The class had come up with norms for the devices ("Put 'em down when we aren't using them," "Mr. Garcia can take them away if we're not respecting them," "Don't lose it.").[5] I jotted down notes, feeling like our research was going somewhere powerful. Students even suppressed the sarcastic eye rolls as I showed them how to send text messages, take pictures, and use the other basic features installed on their devices. The first casualty to theft was still hours away, so my hopes for the research were not yet dampened, and I was excited to see where this collective journey would take us.

Despite all of the research into and enthusiasm about mobile phones, I chose to use iPods rather than iPhones. There were two reasons for this decision. The first was financial: I had a limited research budget that could cover the $199 price tag for iPods and not the cost of pricier iPhones and their monthly usage plans. The second was relevance: the phone part of a mobile device was less interesting and necessary for my purpose of looking at the value of mobile devices in schools. Most people do not use their mobile device very often for making phone calls.[6] Nearly every other popular feature that mobile devices offered could be found on the cheaper alternative. As is discussed later in this chapter, virtually all of the one-to-one mobile technology implementation programs that are taking place in schools globally hand out tablets and laptops, not phones. Further, as I discuss later and as I have looked at in previous work,[7] the meaning of a phone is an important sociocultural factor to consider when engaging with personal devices in schools.

About the Study

So I handed out a bunch of mobile devices, and we did stuff. But it wasn't really as simple as that.

As much as this was a codesigned study—student interests and questions shaped this work just as much as my own questions—I still maintained a responsibility as a teacher to ensure that this work was meeting the academic needs and standards of the students I was legally responsible for. I designed a sequence of learning that relied on the devices.

First, because these devices were a new part of the classroom, our class began by focusing on whom these devices were for, what they did, and how they shaped what can happen in classrooms. This was a week-long inquiry into what mobile devices *do* in schools. Students were able to experiment, ask questions, and collaborate around the kinds of practices that allow these devices to support learning. For me, this was a time for the class to work out the academic and pedagogical kinks of mobile devices together. I also had an ulterior motive: with all of the experimentation, talking, and project-based learning, students were strengthening the classroom community. Relationships were forged in the moments of helping students connect to the school's Internet or add information to a digital mindmap. In considering the conversations that emerged after this initial week, I think that relationships grew stronger largely because playing with mobile devices and, later, playing an alternate reality game stopped feeling like school for a little while. Rather than business-as-usual interactions within classes, this effort at melding in-class research with connected learning put relationships at its center.

Writing, reflecting, and experimenting with mobile devices continued throughout the rest of the class. After the first week, however, the class moved the use of mobile devices to the side as we focused on Ask Anansi (a game described in detail in chapter 4), the novel I had assigned, and various forms of writing and analysis. The class was still an English class, and I wanted students to work on considering how these new devices were augmenting class when they were not the main focus. In short, we were moving the focus away from these devices and toward each other.

Although I encouraged forms of mobile media use within my classroom, my students and I mutually created boundaries and ways to utilize these devices when reading books, discussing poems, and writing (or audio recording) daily reflections. Throughout the time we used the iPods, we sought small areas where these phones could augment traditional classroom structures. Students used their devices to look up information on the California Basic Educational Data System, *Wikipedia*, and the Los Angeles Unified School District "School Report Card" database in order to identify inequities within their school. As students began searching for various resources online and I helped facilitate sites for research, I noticed that Katherine and Holden bookmarked the sites I suggested and sites they encountered through their own online information searches. Holden often pressed the two buttons on the iPod simultaneously to take screenshots of relevant information instead of writing it down. After they told me why they were bookmarking and taking screenshots ("Because it's faster"), I had them explain their strategies to the rest of the class. By sharing student research practices as they emerged—like bookmarking data and taking screenshots of pertinent information—students helped each other invent new critical practices and productivity skills that could be applied regularly in other classes and beyond grade nine in high school. Consultants, book authors, and educational software companies don't necessarily have all of the answers for how these devices should be used. Mobile devices and their pervasive use are largely a new phenomenon. The fact that most schools do not involve students as "consultants" speaks more to issues of trust and assumed expertise than to anything else. When I highlighted my students' innovations, they could see their agency enacted in the classroom, which further strengthened classroom relationships.

Although there are many ways that classroom pedagogies can shift positively when adults acknowledge the cultural impact of mobile devices on youths' lives, mobile media continue to be forced into the marginal spaces of traditional pedagogy. The numerous limitations and disruptions already imposed by schools like SCHS signal that the educational landscape has not yet been able to contend with the very different landscape of learning and engagement in which the world is now immersed.

Curating Youth Research

Before diving into some of the moments of innovation in this project and the many blunders I made throughout the study, I want to touch on why I approach the research and design process collaboratively. The model of civic-focused research I engaged in is called *youth participatory action research* (YPAR).[8] At its most basic, YPAR involves young people in understanding and developing research questions about their world. Working alongside adults, participants in a YPAR project collect data and develop empirical research findings. Unlike some models of research, however, this is not the end goal. Instead, YPAR asks the coresearchers to consider the purpose of research and ask about the outcomes and civic actions that can emerge from the process of researching. For example, in a separate YPAR project with high school students from across Los Angeles, students presented research on issues of school equity to LA policy makers (including their principals and the mayor), who applied some of the students' findings in academic settings. Based on student findings, for instance, the principal at SCHS at the time changed a tardy policy.[9]

YPAR is *not* a curriculum. Throughout the weeks of this study, I wanted students to engage in academic research while thinking critically about the products we were using by asking questions such as "Who profits from the devices and apps we utilize?," "Who has power?," and "How does power change as a result of these devices?"

In the first few weeks of this study, students were invited to see themselves as experts and were encouraged to ask the questions that interested them. In the following weeks, their research and relationships shifted toward a social critique of SCHS and its surrounding community. We used our expertise in creating and questioning to reimagine the school space and the things that it means for its students every day.

Students as Experts

Students are assumed by peers and family members to be experts at using mobile phones, and the data on student use of iPods within my classroom shows that this expertise became evident within the classroom community. A week after handing out the iPods to my ninth-grade English class, I needed to add a PDF and set of audio files to each student's device to support the classroom curriculum. During silent reading, I took my laptop to each student, who plugged his or her iPod into my computer, but an error message appeared on my computer, and I was unable to transfer the files. After watching me growing increasingly frustrated with the computer and iPod, Dante called to me:

> "I can fix it," he said without looking up from his iPod, on which he was typing something.
>
> "It's not accepting my computer or recognizing it, I guess," I said, frustrated.
>
> "Lemme see," Dante said, rising from his seat and walking to the computer.
>
> Taking over the space in front of my computer, Dante asked me which files I needed transferred and quickly showed me how to transfer the files that I had been struggling to add to the iPod.

Although I consider myself technologically literate, I needed to rely on Dante's expertise. And though a passing moment early in this study, such an exchange was important within our classroom's community of practice[10] because Dante's expertise shifted how students in the class understood power. He continued to sit quietly in class, his eyes often glued to his own device, but whenever I had yet another technical difficulty with the iPods, I asked him for his assistance. In brokering our dialogue around mobile devices, Dante became used to me asking him questions, engaging him in dialogue, and bringing him directly into classroom conversations. (In chapter 4, I dive more fully into Dante's journey as a digital expert and classroom leader.) When devices that are a central part of youth socialization are used in the classroom, benefits can come from relying on youth expertise and thus dispersing power. Having Dante assist me brought him more fully into the classroom environment and allowed me to learn from the knowledge my students possessed long before they walked into SCHS. When focusing on academic content, I often assumed the role as classroom expert to begin our work. However, when the class was troubleshooting the school's wireless network and iPods or investigating historical inequalities within our school, student expertise guided the classroom work.

One aspect of using these devices that students were not familiar with involved having them create quick response (QR) codes to communicate and participate in a classwide game. QR codes are square, barcode-like pieces of text occasionally printed on products and advertisements. The general public doesn't seem to use them regularly in the *real* world, but they are easy to create and read with mobile devices and became a useful tool in our class. QR codes are essentially a personalized text and are a productive tool like those that grant users the ability to create a text message, an essay, or a photograph.

The QR codes for this unit added feelings of exclusivity and strengthened the class community. As I printed out and posted on the classroom walls the numerous QR codes students sent to me in response to daily quickwrites and class discussions, our walls became encoded with the sentiments and frustrations of the class: "I miss my dad," "[Jay] is the best at dis," and "Its not fair when they call home when Im late to school." These are examples of QR codes that proliferated in the class and that students scanned with their iPods to decode. As I prepared before the class bell rang,

Figure 3.2
A student's QR code for "I miss my dad"

I could see Tess and Marjane scanning codes and nodding knowingly. The codes helped illustrate the ease with which students created and deciphered QR codes, and they also allowed the humanizing sentiments of being a student at SCHS to be shared and accepted by others within the class. Even though QR codes are used as marketing tools in mainstream media advertisements, magazine covers, and even business cards, the students in my class and in the focus groups I conducted did not recognize them or know what they were called. They seemed like a secret language that my students were able to detect, create, and use for dialogue with their classmates. Like a digital pig Latin, QR codes allowed students to communicate with each other, with me, and with those who found their clues in a way that was occluded from the dominant public gaze.

Wresting Away Adult Authority via Youth Expertise

Like computers today, iPods allow users to set a password to deter unwanted users from accessing and deleting content. Additionally, users can (wirelessly or via a USB plug into a computer) load their iPods with applications that customize each device based on their users' viewing, listening, playing, and productive dispositions. Prior to handing out the seventeen iPods to my ninth graders, I created accounts for each student using my own computer, installed several applications on the devices that I anticipated we would use in the classroom, and used my own administrative password to install additional applications, music, and other media content on the devices. These additional programs included a text-messaging app, a mind-mapping utility, and the *Wikipedia* app. Even as I set the administrative password to control the devices, I recognized that this severely limited the social practices that these devices richly enable. In fact, in work I did after concluding this study, it became clear that the meaning of a phone is lost when it is no longer tied to personal identity and control.[11]

As a teacher, my principle concern was the learning that took place in my classroom. As much as I attempted to build community, empathy, and trust as core values in the classroom, my emphasis on technology blinded me to the potential repercussions of locking the devices. Even though I attempted to cordon off how mobile devices could be used, I installed ways for students to communicate socially with each other, including the text-messaging app and the popular Facebook app. Further, I acknowledged to my students that they were likely to use their own mobile media devices to install applications and "be social," but I noted that my intentions were to apply these devices as tools for shaping pedagogy and learning. I also reminded students that this was an experiment we were conducting together, so their thoughts on learning with mobile devices would be important for the class. However, again, myopic assumptions—that two phones are better than one or that having a "work" phone and a "social" phone would work for students—clearly thwarted the personalized value of having personalized devices in classrooms. Holden's protection of his *real* phone in this chapter's opening is perhaps the clearest indication of this myopia.

This acknowledgment leads to one of the major limitations of this study. By providing students with an additional, teacher-provided device, I collapsed the natural familiarity that students have with phones and ended up making them an impersonal imposition. It would have been much more natural for students to use their *own* mobile media devices during class activities because the devices would be familiar to them from a technical proficiency standpoint and also ascribed with personal value. Some secondary schools and universities have embraced a bring your own device (BYOD) model for proficiency, personal value, and economic reasons. Additionally, my students were informed that if they created a password on their iPods to limit others' use of the devices, they needed to share the password with me so that I could administer and add information relevant to the class. Again, this imposition of adult authority is a break from the personal meaning placed on these devices. I intended to control how students used mobile media in classroom contexts. However, within a day of handing out the iPods, I realized that I had (of course) underestimated the ability of my students to overcome my digital mandate.

The day after handing out the seventeen iPods, I noticed students coming into class transfixed by games, music, and customized backgrounds on their new mobile devices. Within this twenty-four-hour period, students had figured out how to install applications on the iPods, including the game Angry Birds, pirated music, and even a program that acted as a local police scanner. Students coopted and personalized their mobile devices in ways that weaved social value with academic value. Of course this is what students would want to do with their phones. This was a valuable step toward ensuring these devices remained a relevant part of their lives. And yet I initially saw these adaptations as a security failure: I hadn't controlled these devices enough. I had not yet realized that in bringing mobile devices into the classroom for academic purposes, my biggest oversight was not planning for students to include social uses for them.

Personalization and Ownership

When my students broke through the passwords I'd established in the classroom, they were participating in an extension of the usual kinds of cat-and-mouse authority challenges that play out with mobile devices in schools. Prior to handing out these devices, my students frequently hid phones and iPods beneath their desks or in their backpacks when using them during class. It is the type of behavior—discrete reading and sending of text messages and swiping in the middle of class—that I now see carried out in my university classes. And again, I've been guilty of such behavior, too, and have sent a quick response to a query while sitting in a meeting. The subterfuge with which we engage with our mobile devices has become a common (and problematic) part of adult culture. After my students received their new iPods, they flaunted the latest additions and modifications they made to the devices on loan to them. For instance, during the second week of class, Ras showed me the background image he had installed on his iPod.

The fact that Ras demonstrated his new background while also showing me the pages of games and applications he installed suggests that he saw the iPod not solely as an educational tool but also as a portal for entertainment and identity construction. During the thirty-minute advisory class that I taught, Ras sometimes allowed other students to use his iPod to play games. As Ras sat at his desk turning the pages of a book or talking with a classmate, he seemed pleased that his device was being used by others. By borrowing and using his device, peers accepted Ras's iPod as a demonstration of his curatorial choices. Watching Ras circulate his borrowed iPod, I was reminded that the content that he and other youth curate on their mobile devices does not simply reflect personal tastes but also caters to peers for acceptance, socialization, and affiliation. Ras's screen demonstrates a preference for a colorful backdrop for his device and a predilection for gaming, but it also signals his familiarity with mobile-device use. By looking at his screen, peers can glean that Ras can quickly and—based

Figure 3.3
Ras's customized iPod screen

on the titles of many of the displayed apps—freely add content to his phone.

The image of Ras's iPod also highlights an important theme within the data I collected during this study. Students personalized their devices in significantly different ways. As I kept an eye on student's iPods over the weeks of the study, I noticed that nearly every device in the class had been personalized. Students created screen backdrops, placed the iPods in cases that glittered, and used bright green earbuds instead of the standard white earbuds provided with the devices. Their devices never looked quite like those that I initially handed to them. Although some students, like Dede, did not add additional programs, their iPods were still customized with images that were imbued with personal meaning and value. Dede's iPod background showed a stock image of a rose, which personalized her device. Even as a tool that was not her primary means of socialization or communication, Dede's iPod was personalized so that it signaled to classmates aspects of her identity.

Dede, Ras, and the rest of my students took personal ownership over the appearance of their mobile media devices and the content included on them. They illustrated to me that by ceding adult authority over mobile media use in my classroom, even involuntarily in this instance, student voice and student identity were naturally included within the classroom. Although not all students were as eager to share their iPod modifications with me as Ras was, they comfortably displayed and shared their content choices with each other in the classroom community. As students shuffled into the room before class began, they exchanged tips on where to download specific images, which songs they were listening to, or how to beat levels of certain games. In this way, the social talk between students about their iPods seeped into the participation in the class.

There are two considerations here for schools and educational policy. First, by stifling student use of mobile media, many teachers also cut off students from the social networks within their communities and from outside resources made available on Internet-enabled devices. They essentially remove the cultural value of mobile devices and coopt them for academic purposes.[12] Second, stifling student use of mobile media also stifles how students express and behave within a classroom setting. As I continued to look at how students "act out" when texting or utilizing mobile devices, I saw these actions primarily as efforts for students to demonstrate aspects of their identities. By displaying something as innocuous as a digital flower on the screen of an iPod, students imbued their school space with personal meaning and value.

Entertainment Portals

In this chapter, I am owning up to some of my mistakes, and I now want to look at the elephant in the mobile media room. Although I initially hoped to present a picture of my mobile-engaged classroom as a utopian, distraction-free space, that simply was not the case. It was *never* the case, even before I handed out iPods to my students. Subterfuge with devices has become a regular part of life at Los Angeles high schools and schools across the country. One notable exception was New York City, where until recently[13] mobile devices were prohibited in public schools. This curtailed distractions but also created avenues of profit on the backs of students, who paid daily fees to keep their devices in "cell-storing trucks" while they attended classes.[14]

I would have preferred that my students power up their iPods solely to guide them toward continually engaged, focused learning experiences. However, the devices often impeded learning and focus in my classroom. In attempting to note these incidents while also teaching the class, I tallied times that distractions from student devices affected the classroom environment. I wrote down times when I noticed students were focused on using their iPods in ways that were not tied directly to the classroom activities, times I asked students to stop using devices, and moments when students talked with one another about content not related to classroom material. Additionally, each day I reviewed my audio recordings of these classes to see if there were lapses in my notes about these instances of distraction. Were there times when I was frustrated by the lack of focus in the classroom? Of course. I clearly faltered at times as an educator and felt frustrated with how technology use in the classroom distracted my students. Over the course of seven weeks, I tallied 138 instances of mobile media as sources of distraction. This included both the iPods I provided and students' personal phones. This averages nearly four instances of class disruption per day for one class period.

Digging into these coded instances of distraction in my classroom, I noted that there were three general categories of disruption—socializing and communicating, listening to music, and gaming. In addition to noting the types of disruption, I also tallied times I intervened and times I let disruptions go unaddressed. Corroborating the findings discussed in the previous chapter, most "disruptive" student mobile use in my class was for social purposes (82 of the 138 coded instances). Additionally, my efforts to stifle these distractions by intervening generally had mixed results. Between 38 and 50 percent of the time, students continued to use their mobile devices even after I asked them to stop.

Although this data is specific to my own classroom, it suggests that my attempts at reigning in mobile-device use for academic purposes found resistance in the student interpretation of school time as fluidly social and academic. Looking at both student behaviors and my own actions in the classroom, the patterns of disruption and adult intervention reflect blind spots in classrooms. Even though my study of the entire SCHS campus confirmed that teachers stifled students' primary purpose for using their devices (to socialize with one another), I ended up replicating this same process in my own classroom.

Minerva and Nondigital Production

If we're going to embrace connected learning and wireless classrooms, we need to think critically about what this means and in whom we invest our attention. Much of this chapter highlights the things that students *do* with mobile devices, but *nondigital* tools also support student learning, relationships fostered in classrooms, and the fundamental civic purpose of public schooling. With this in mind, I want to share a vignette of a moment of classroom work that was almost entirely divorced from the digital devices that started my inquiry.

After my class and I moved beyond the typical enchantment with new devices, there were moments when digitally connected youth produced multimodal content, collaborated with peers face-to-face, and generally learned *wirelessly*. Below, Minerva illustrates that a culture of participation does not always rely on tools that blink, flash, and need to be recharged.

Throughout the year, I saw Minerva in the halls of the ninth-grade academy building during my conference period. Frequently, she'd inform me she was in the halls because she was kicked out of class. Her language and decisions not to participate in classes contributed to these frequent hallways appearances. In group discussions in my own class, Minerva regularly weaved together curse words with academic language. Her mastery at this profanity-laden vernacular impressed me but significantly limited her academic growth by frequently getting her kicked out of other classes. In her first two quarters of high school, Minerva failed most of her classes. She earned a C in my English class and in her social studies class. In the vignette that follows, Minerva and I discuss ways for her to engage with media deliberately and personally. Highly personal, Minerva's text and her initial hesitancy in producing it are an example of how classroom innovations often come not from charging and utilizing electronic devices but from investing in trust and relationships:

"I've got stories to share, but I think they are too much for kids."

This is what Minerva tells me in class after I ask students to continue their work. She is quiet and introspective as she often is when she and I speak quietly by my desk or in the hallway. As other students work on developing an autobiographical children's story, Minerva brushes the dark hair out of her eyes, which are welling with tears. Although this is one of the few assignments in the quarter that I've asked students to complete with paper and colored markers, Minerva's response to producing this media product is similar to other assignments I've asked her to complete in the class—reluctance coupled with personal investment in producing authentic and meaningful work. If Minerva is going to complete this assignment (and at this moment, it's a real "if"), she is going to need a real audience that can learn from her. Minerva's engagement with producing media products related to her research topic ("the absence of love in South Central Los Angeles") has resulted in highly emotional reflections on her life.

"So you mean my story doesn't have to be a happy one?" Minerva asks.

I respond, "No, it doesn't have to be happy—if you think that's the best way to convey your story and your experience with the research theme you've investigated. Of course, there are happy children's stories, but this is an opportunity for you to share the knowledge you have in a way that would be understandable for young people."

As Minerva and I discuss the powerful role and responsibility of being a media producer, I recall two sentences she wrote while I guided students through the process of producing quick response (QR) codes. During one of the lessons on using the QR app in the class, students were asked to write any single sentence, create a QR code with the sentence embedded within it, and send it to me. Minerva's two sentences were "I hate school" and "I miss my dad."

Minerva shuffles back to her seat, folds several sheets of paper in half to make the pages of a book, and begins to write.

Pausing occasionally to punch responses to incoming text messages, Minerva completes a draft of her story, which she hands me with a cheerful, "Here you go," before bouncing out of class for lunch. Each page has text carefully written in only the bottom third of the paper, saving space for forthcoming images. There are no crossed out words. Although the draft has some grammatical and spelling errors, it appears to me as though it is written in a single coherent thought from beginning to end:

Once upon a time there was a girl named Jessica. She loved her mom with all of her heart. They were so close like pb&j. They were like best friends, sisters, anything you can name they told each other everything good or bad and always stuck by each other. What a good mother.

So one day, little Jessica was in the park enjoying her day with her love, then she gets home everything falls apart in her life. She can already feel it. She got some really bad news about someone she cared about so much. She can feel her heart fall all the way down to her toes.

Well guess what her mother had got into some trouble with the cops and the cops went crazy looking for her. One day, they decide to break into our house. They throw everyone on the floor and handcuffed us. What a terrible night to remember.

So the very next day Jessica's mother finds out about what happened and she decides to take off and that day Jessica dropped so many tears, she felt like she had lost everything she ever had. No one could make her feel better. She had lost her mother.

Ever since that day she feels really lonely especially when she sees mothers and daughters together. All Jessica does is think about the past and every moment she spent with her mother. But deep inside her heart she knows her mother is by her side.

What a sad story. So yeah, hopefully they could reunite one day but for now that's about it. We should all enjoy every single day with our loved ones because you never know what can happen. The end.

Unlike many of the other narratives constructed by her peers, Minerva focused on issues of loss. In producing her own narrative for an audience of younger children, Minerva illustrated a history of pain without consolation. The way that Minerva's story was processed and reinterpreted as children's literature pointed toward her understanding of both literary structure and interior turmoil. The line between fiction and nonfiction was intentionally blurred in her story. There is regret and anger in her story—not at her mother but at the exterior tensions that crashed in on the private sphere of family life.

There is an unintentional slip in Minerva's narrative that was revised in future drafts. The story begins as a third-person tale about "little Jessica" and her mother later and becomes a first-person account of a negative encounter with the police: "they decide to break into our house. They throw everyone on the floor and handcuffed us." By the next page, Minerva resettles the narrative back into the comfortably distanced space of a third-person point of view. The moment at the crux of this story that ends a childhood of happiness is highlighted by this narrative slip-up from third person to first and back again, like the bumping of a record player: it pulls readers uncomfortably out of one place to reinterpret the narrative.

In most children's books, there is a lesson or moral, and Minerva offers a trite statement at the story's conclusion: "We should all enjoy every single day with our loved ones because you never know what can happen." But there is a deeper insight to be gained from this story. Although this final paragraph suggests that the story is about losing a loved one, Minerva's text predominately critiques ways that she is powerless before the forces that affect her life (such as the police and "a terrible night to remember"). In Minerva's narrative, family and happiness are deracinated in ways that adhere to children's story tropes. In her advice to readers about enjoying every day with our loved ones because we never know what can happen, Minerva shares guidance for a world that she distrusts. She makes the familiar environment of home and community unfamiliar as she invites readers to shuffle through the corridors of her memory. There is no warmth in these final sentences of advice. This is the conclusion of a story about separation, loneliness, and estrangement.

It is worth considering how digital tools could enhance Minerva's media production here and how they could get in the way. On the one hand, Minerva's powerful narrative could be amplified for peers, youth, and the public at large, acting as a powerful familial narrative within South Central High School. This story could be linked to, tweeted, and shared. However, complicating the potential benefits of media production with a culture of participation, this work may not have been produced if it were written for a public audience that Minerva did not necessarily trust. What moments of trust and community building do we lose by assuming that things are better and made more complex when they are typed and tweeted?

Weeks after Minerva turned in her story draft, I revisited my transcription of one of our final conversations during the study. Sitting across from each other a few minutes before the bell rang for lunch, I asked Minerva, "Do you think you've changed how you see your role in the school?"

Minerva turned the answer away from herself and gave a response that discussed the ways games, technology, and research affect students in general: "Now, you can start talking about it, telling people about it. You can just point things out more." She paused for a moment, inspected a cuticle, and said, "You're a teacher that understands us. … I don't know. Probably I got so much done because you understand me. Before you even kick someone out, you try and understand."

Building civic-focused instruction has to start with building trust and supporting classrooms as safe spaces to exhibit vulnerability—for both teachers and students alike. Storytelling can function as a tool for student agency and as a means for students to speak about lived experiences in ways that empower social participation in the future, but these tools must be coupled with the cultivation of community, trust, and safe spaces in schools. By writing a children's story that describes her anger and sadness about how her family was broken apart, Minerva claimed a space for action and advocacy.

Minerva's narrative highlights how classroom trust can support student learning, but I don't think that trust and the student expertise afforded by mobile devices are mutually exclusive. A classroom that utilizes mobile devices can (and should) be a humanizing classroom, too. This story came out of a class that began with a warts-and-all inquiry into mobile devices, and Minerva's personal voice found strength and courage through the opportunities that these digital tools encourage. When we think of wireless classrooms of the present and of the future, we need to consider how trust and relationships work in tandem with and are supported by devices—not the other way around.

Conclusion: About Disruption

For teachers, the word *disruption* signals a problem within the classroom. The ways I have used the word throughout this chapter illustrates what all teachers know: disruption is a major problem in classes and needs to be avoided at all costs. Entire teacher education courses are shaped around "managing" classroom disruptions. However, outside of the classroom, when people discuss developments in technology, a disruption often is seen as positive. Startup firms often look at ways to disrupt existing markets. Uber and Lyft have disrupted the taxi business, Airbnb the vacation housing market, and online streaming services like Spotify and Apple Music the online downloading business. A disruptive device—like a smartphone or an iPod—changes how people interact, behave, and produce work. Disruption reframes how knowledge is shared and produced. In this sense, globalized economies are disrupted as a result of advances in digital technology.

Nearly everyone today is interacting or working in ways that have been disrupted by mobile devices—everyone except people who spend time in public schools.

When we invest in technology in schools, we do so in rudimentary ways: we pay a lot of money for devices that we hope will fix the persistent problems of public schooling. Minerva reminds us that wireless classrooms need to be about trust and expertise. I hope my blunders here remind us that teachers are fallible participants in classrooms and need to work alongside young people. Maybe investment in disruption should look to the people in classrooms rather than the devices.

In discussing the kinds of successful attributes we see young people exhibiting today, educational researcher James Paul Gee describes them as "shape-shifting portfolio people."[15] With a suite of demonstrable skills (that can be displayed online in a portfolio), Gee sees young people use skills in one context and then quickly switch to use those skills in another context. This seamless "shape-shifting" means that today's young people might work in varied industries and leave behind traditional models of labor where workers toiled in a single profession for decades. Mobile devices, as we see in this chapter, disrupt and help with this shape-shifting process. They bisect the traditional places where text is consumed, produced, and analyzed. They make the place of learning even more contextual. Likewise, communication now can be disrupted by both place and time. My students texted their class updates to me even when I was away from the school building and a substitute was teaching the class (as is detailed in chapter 5). Similarly, my experiences using social networks like MySpace and Facebook in my classroom[16] have shown me that conversations do not need to happen in "real" time. A conversation on a space like Facebook can continue through comments and updates over weeks, offering new narrative modes of discussion and means for online public performance and privacy.[17]

Traditionally, school-based learning has been anchored to classrooms, desks, and specific uses of academic space. By documenting lived realities, facilitating discussion, and allowing for collaboration with individuals both in and out of schools, mobile devices help teachers and students to recontextualize the contexts of school-based learning. How space is read and how space changes learning[18] are key areas of mobile-device use for schools (as is shown in the following chapter). I am also wary that this attribute of mobile phones can frustrate how educators "manage" classroom discipline.

Although I believe that the disruptions of mobile devices can be funneled into strategies to reshape classroom practices, I also know that schools can't have it both ways. If we want innovative classrooms that incorporate the affordances of technology, we have to acknowledge that the square pegs of authentic uses of such technology cannot fit easily into the round holes of existing schools.

When I look back at how I communicated with students, I'm reminded of how my experiences as a teacher evolved. For instance, although Katherine was usually a quiet student when she was in the classroom, she sent me contributions almost daily via e-mails and text messages. Some of these were received during the daytime hours when the class was held, but I also received e-mails of her work in the evening. For instance, during the third week of the class, I asked students to reflect on how outsiders might view see the students' South Central community. I received an e-mail from Katherine that begins as follows:

> May 19, homework; I think our community has a bad rep. Because we kinda make it look bad but unintentionally. & then we realize that people judge our community because of us but in reality we can change this because we're not really bad as people we're just human beings. I could't really get the pictures because idk [I don't know] what proves my statement

Although Katherine points to the shortcomings of her work, her comments allowed for an opening dialogue the next day in class. I asked Katherine if she would mind if I shared her sentiments with the rest of the students, and her homework became the lynchpin for an ongoing discussion about perceptions of South Central Los Angeles. In this case, a quiet student like Katherine, whom I often gently prodded to participate, emerged as a natural leader. The iPods expanded the engagement options available for students for submitting work, voicing ideas, and engaging in multimodal inquiries. Katherine flourished as a student in the hours after school. In her own home and at times that were convenient for her, Katherine's traditional expectations of learning were disrupted.

4 Ask Anansi: Informing, Performing, Transforming

As I handed out the papers that described the location of the first clue, students took the sheets with their standard nonchalance for schoolwork. I tried to make the class interesting, but for students, this was just another activity in another class—another worksheet that stood between them and lunch. I watched as they quickly scanned the paper in front of them:

This badge is tucked away in the corner of one of Mr. Garcia's favorite things: a book! There are many here, so which one is it?

To successfully find this badge you need to find a book someone wrote about his own life, just like you have. He had many names: Homeboy, Satan, Minister, El Hajj Malik El Shabazz. This badge can be yours if you look on page 165.

As soon as Dante realized that he was looking for something in the classroom and that this was a game, he stood up and started pacing, looking for things that might be askew on the untidy bookcases. Many students looked uncertainly back and forth from their paper to me for further instructions, but Dante roused Solomon and Jay, and all three soon congregated around the bookcase.

As Dante and his friends looked through my shelves of books, Tess focused on the final sentence of the clue description. She pulled out her iPod, placed the device in the middle of her table to collaborate with Holden, and typed "el hajj malik el shabazz" into her *Wikipedia* mobile application. The result redirected users to the page for Malcolm X. While more students congregated around the bookshelves, Tess sat and took stock of our room. To her left, nearly half of the class was scouring the bookshelf. Scattered in front of her, the rest of the class, including myself, was looking at the scavenger hunt clue or at the mass by the bookshelves and the walls cluttered with student work.

And then she saw it.

Perched high above the white board, strikingly out of place once it was noticed, a single black paperback leaned against the wall. She nodded to Holden, and they slowly rose from their desks attracting nobody else's attention. She dragged her chair to the shelf, and Holden stood on it, retrieved the black book, and handed it down to Tess. Flipping to page 165, she saw a QR code paper clipped above an underlined passage and looked at Holden, who yelled, "We found it!" Surprised "You did?" comments (and disappointed grumbling from Dante) came from across the room.

Tess explained how she and Holden found the clue and read the underlined passage aloud for the class:

> The devil white man cut these black people off from all knowledge of their own kind, and cut them off from any knowledge of their own language, religion, and past culture, until the black man in America was the earth's only race of people who had absolutely no knowledge of his true identity.

I was prepared to dive into an engaging conversation about Malcolm's words, but the class had other plans. Before anyone else could speak or reflect, Minerva quickly said, "That's a nice quote. Okay, what's the next clue?" The rest of the class looked to me expectantly.

Although Dante seemed disappointed that he didn't find the book, the class as a whole was interested in the process of looking for and eventually finding a tangible object. This class experience was different from what students were used to. For several weeks, my students ran around the SCHS campus (yes, often literally running), hiding clues, snapping pictures, and clandestinely placing notecards on various parts of the campus space. They also had in-depth conversations with a mythological figure, Anansi—a West African trickster spider—and tried to determine who was the "winner" in our class at any given time. This kind of English class wasn't what school usually was like.

In this chapter, we look at how games helped my classroom become a transformative space that was guided by student inquiry and sometimes facilitated by digital tools. It was fun, and moments of powerful learning resulted from the classwide game. As I note in chapter 3, I made plenty of mistakes as I rolled out this game, and I am going to share more of those, too. Just as a vocal group of supporters praises the life-changing potential of technology in schools, another group of enthusiastic pundits promotes game-based learning. In sharing what was done in this study, I might be able to temper some unrealistic expectations. But in order for me to explain the value of having students in my ninth-grade English class run around and talk with fictitious characters, I first need to discuss the roles played by games in schools and the ways I tried to blur reality and fiction in my own classroom.

Why Games?

There is a long history of educational games that try to improve student learning. In particular, the value of digital games in helping kids to learn has long been publicly debated. Playing digital games—getting lost on the Oregon Trail, chasing after Carmen San Diego, collaborating in online virtual worlds like Minecraft and World of Warcraft—is a significant part of how new technologies can be harnessed in schools.

Recent research highlights the academic, civic, and creative potentials of video games,[1] but I have been interested in how the concept of play more broadly allows for students to learn in new ways. When we unplug play from digital contexts, we can imagine classrooms where students take on playful identities, create, explore, and challenge one another in ways of engagement that are significantly different from what schools have historically offered. Nearly a century ago, for instance, educational philosopher John Dewey wrote that, "From a very early age … there is no distinction of exclusive periods of play activity and work activity, but only one of emphasis."[2] Dewey also describes games and play as "relief from the tedium and strain of 'regular' school work."

What differentiates work and play? Play and games? How is energy dispersed and focused, and how does a person's agency guide and make decisions within work and play? Traditional models of schooling emphasize work and mental exertion in contrast to play, but games offer opportunities for students to escape from such stifling environments. Games reconstruct the possibilities of learning and the ways learning is felt by the students who are immersed in it.

Going a bit further, there are many different definitions of games and subdivisions of what happens within games.[3] For the purposes of this chapter, I rely on Katie Salen and Eric Zimmerman's definition—"A *game* is a system in which players engage in an artificial conflict, defined by rules, that results in a quantifiable outcome"[4]—because it highlights the transformative potential of many games. They point out that games can act as a portal toward social change. Specifically, *transformative social play* can shape learning opportunities that incite students to wonder and can expand student learning to include engaging with the larger world. Salen and Zimmerman note that such a model of transformative social play

> forces us to reevaluate a formal understanding of rules as fixed, unambiguous, and omnipotently authoritative. In any kind of transformative play, game structures come into question and are reshaped by player action. In transformative social play, the mechanisms and effects of these transformations occur on a social level.[5]

The other fundamental concept around gaming that is important to consider for schools is the idea of a "magic circle." Illustrated decades ago by philosopher Johan Huizinga, long before electronic gaming and electronic mobile media existed, the "magic circle" reminds us that when we play a game, we are "stepping out of 'real' life into a temporary sphere of activity."[6] For instance, when people play charades, they accept the silliness of acting out silent clues and having other players guess and yell out words frantically because such behavior is within the "magic circle" of a game. Behavior that is not considered socially acceptable is allowed within a magic circle because participants and viewers know that this is a temporary space outside of social boundaries. In the "magic circle" of game play, role playing, acting, and behavior can ignore social norms, and students can pose, flex, and experiment safely. A "magic circle" gives players a safe space in which to learn and be vulnerable.

The general notion of the magic circle is largely in opposition to general assumptions about the purpose and operation of schools. Schools provide spaces that model "real" life, and some behaviors—including conforming to society's power norms, adhering to bells that dictate a factory-like schedule, and acting politely—are parts of the "hidden curriculum" of what kids learn in schools. In contrast, the magic circle allows wild experimentation and behavior because gameplay is a safe space to be someone that society says you are not. To me, the revolutionary potential of gaming within classrooms is not the fancy technology or structure of a game but the radical possibilities for trying on new ways of being and for encouraging youth to imagine identities that are unconstrained by local norms. Games can create vastly alternate realities.

An Alternate Reality

As hinted above, one contemporary genre of gaming that I've found to be a powerful model in my classroom is something called an alternate reality game (ARG).[7] An alternate reality game blurs the notion of the magic circle by having a fictitious narrative take place *in the real world*. This popular genre has mixed how participants interact with media and their everyday world by sending players to find ringing payphones in public,[8] conduct protests over "fake" topics,[9] and even sneak into car dealerships to locate a stolen vehicle.[10] Most players know that the game is built around a fiction but regularly pretend that "this is not a game" (a popular trope of ARGs).[11] Blurring what is real with what is a game is largely about immersion and, to a degree, challenges the notion of a magic circle. Based on player reactions to ARGs, it seems that pretending that "this is not a game" makes the experience of playing them feel more fun.[12]

In her research on the ARG I Love Bees, game designer and researcher Jane McGonigal notes how thousands of individual players worked together like a "hive" to unravel a series of complex puzzles.[13] Building off of a theory of "collective intelligence," players developed a social network of expertise and puzzle solving.[14] Thinking about the capacity of such robust learning in ARGs, I am reminded of McGonigal's comments in her 2010 TED talk, where she notes that "gamers ... are a human resource that we can use to do real-world work ... that games are a powerful platform for change."[15]

Taking McGonigal's call about the possibility of games seriously, years before the study described in this book, I collaborated in designing an ARG for my classroom focused on ecoliteracies and youth civic engagement.[16] Briefly, the Black Cloud ARG imagined that the pollution in Los Angeles was so bad that a cloud had developed consciousness and was communicating with my students via Twitter. Utilizing sensors that measured the air quality in hyperlocal spaces, students assessed the air quality in and around their school community, critically examined this data, and developed environmental recommendations that were presented publicly. In postgame interviews and reflections, student data identified specific ways that this game affected student literacies and environmental awareness. After acting out the roles of "citizen scientists," students informed their families and friends of air pollutants around them and advocated for school environmental improvements.[17]

The experience of developing and executing the Black Cloud ARG illustrated to me that play, the blurring of fiction and reality, and the ability for students to take ownership over parts of their community were salient factors in student motivation. This experience ultimately led me to Ask Anansi.

Ask Anansi

Building off of my previous experiences utilizing games and ARGs in the classroom, I designed a game called Ask Anansi as an experience that wove together participatory culture, technology use, and classroom socialization. Rather than being a game played *on* mobile media devices, the game was designed so that it was facilitated by technology and largely driven by a narrative.

Briefly, a West African folklore character called Anansi serves as a "trickster" figure in Caribbean tales.[18] Taking the guise of a spider, Anansi often deceives others, and his cunning illustrates crucial cultural perspectives. In one telling of an Anansi story, the spider ends up the keeper of all stories. From this auspicious beginning—a trickster spider that hordes the truths of the world—Ask Anansi was created.

In ARGs, players are part of and come to understand a larger story. They experience it and shape it. Below, I share the story that I designed for students to experience and the branches of game play that affected student learning. These narrative and pedagogical decisions were developed for an ideal world, but how Ask Anansi actually unfolded comprises the remainder of this chapter. As a game focused on the needs and experiences of the ninth graders in my classroom and my own knowledge of our classroom community, I did not design Ask Anansi as a replicable experience. However, I offer guidelines for teacher-driven game design and the original concept document for Ask Anansi in appendix B.

The Anansi Story

Shortly after my ninth-grade English students received their iPods for this study, they were contacted by "Anansi," who sent them cryptic e-mails and text messages. Although students did not know who Anansi was, they did in-class research into the mythical figure and had questions about what his trickster motives might be.

After a classwide rapport was developed between Anansi and students, he offered to take requests for stories about South Central that students would be interested in hearing. However, as a trickster, he did not provide an immediate and fulfilling story to students and instead elicited information from the students. He asked them for pictures, interviews, primary document evidence, and their own opinions about the things they wanted to know. In essence, he *tricked* students into conducting their own inquiry projects.

While students were working alongside this trickster, Anansi also started leaving physical clues on campus about his current whereabouts. After receiving clues, students searched the campus for Anansi's notes. Passing his tests, the students were recruited to help Anansi further rename and retell stories about the school and community. Working side by side, students and an imaginary spider spun webs of new perspectives on the world around them.

The Gaming Sequence

Three key sequences of play unfolded in the Ask Anansi alternate reality game. First, students used mobile media to communicate with and learn about Anansi's motives. Second, after students were in dialogue with Anansi and thinking about the local stories of their school and community, Ask Anansi invited students to imagine a question they would like to answer and then to create critical narratives of their own around these questions. As students shared their findings with the class and with Anansi, he asked for refinements to questions and answers, requested additional data, and posed additional challenges. In effect, Ask Anansi functioned like an inquiry engine. Students conducted research in iterative cycles that began with big questions—"Where do stereotypes come from in South Central?" and "Why is there an absence of love in South Central?"—and delved into more specific and personal questions. Additional topics for research included graffiti, violence, and poor-quality food at SCHS. I intentionally portrayed Anansi as an annoying character who always kept the truth out of the grasp of students. The conceit of mischievous trickery was a sometimes silly one, but students took the opportunity to seek truths about their lived experiences seriously.

The final phase of the game was a multiweek scavenger hunt conducted on campus. Anansi modeled the mechanics of how to search for and participate in the hunt (as portrayed in the opening of this chapter). After students investigated the campus through these clues, they were recruited by the spider to be "Agents of Anansi" and took ownership of their own stories and the spaces they inhabited. Anansi guided students through the mechanics of finding hidden badges throughout the school, making their own clues, and ultimately, using these skills to tell their own powerful stories. A weekend-long extended hunt for dozens of clues was a thematic climax for the game. However, the students' work was far from done. Students used the scavenger hunt game model as a mechanism for sharing their inquiry research with classmates. In fact, during the final phase of Ask Anansi, students wrote museum-like placards for the locations that were previously "hidden" as scavenger hunt locations. These cards publicly shared the student counternarratives with anybody who encountered them in their daily interactions at SCHS.

With the first ARG I codeveloped for my students, the Black Cloud, some students felt deceived by my statements that "This is not a game." So Ask Anansi was framed as a game, and many students quickly assumed (correctly) that I was the one who was e-mailing and texting them, pretending to be Anansi. With this conceit unspoken but largely acknowledged by the class, this kind of pretending was accepted within the class, and students responded to and communicated with Anansi.

Different Games for Different People

Ask Anansi folded together game play, research, and narrative-driven information. Students participated in a scavenger hunt to share their research data and to appease an increasingly frustrating and elusive spider. Like my experiences with the Black Cloud, students took different aspects of the game more seriously than others. For some students like Dante (profiled later in this chapter), the competitive nature of a game was a significant motivator for participation and learning. For some students, the story of this spider interjecting himself into the class was intriguing, and the narrative of the game was a highlight. Some students appreciated the opportunity to drive the research topics and to gather and share the difficult questions they asked.

Ask Anansi functioned differently for different students. Because there are no magic bullets in education, we can't assume that one game is going to be the cure-all for engaging every student. Instead, I designed this ARG to have multiple points of interest for my students.

Students were in dialogue with an imaginary spider and well on their way to interviewing, documenting, and researching their research questions. The remainder of this chapter focuses on the scavenger hunt sequence of this game. Stripping away much of the ARG narrative, I look at moments of play and decision making by students. Extending the Ask Anansi fiction into the realm of teacher professional development, I have included in appendix C an in-depth interview with Anansi about the role that teachers play in educational game design.

Inform, Perform, Transform

Throughout much of the rest of this chapter, I tell the story of what happened when students used an alternate reality game to play and tell stories at SCHS. The game Ask Anansi certainly got my students out of their seats and playing in school, and they experienced academic and civic learning when playing this game. In particular, I want to discuss how students used this game to write counternarratives about their community and its existing dominant stereotypes.

After back-and-forth dialogues with Anansi and after individualized research on their own inquiry questions, students had gathered enough information and data to be able to join a larger public discourse. Rather than simply accept the stories and assumptions that outsiders frequently made about South Central, students participating in the process were empowered to tell their own stories and define what SCHS and the South Central neighborhood meant to them. All of this was done within the conceit of a game and its narrative. As the game designer, I wanted to support students' problem-posing critical thinking and civic participation, but the narrative goal of this ARG was for students to satisfy the insatiable need of Anansi for a good story. After students critiqued the existing dominant stories related to the topics they chose—trash, violence, graffiti, pollution, stereotypes, love—they retold their own version of the story that they were offering to Anansi. A retelling of violence in South Central, for example, was informed by students' own knowledge, interviews with family and community members, a historical analysis of the geographical and social space, and an analysis of dominant media portrayals. After this written work, students moved from analyzing and writing to proposing ways to shift these local narratives in the future. At this stage, the class embarked on a multiday scavenger hunt sequence that illustrates how students can positively hack and recontextualize their physical surroundings.

Building off my previous experiences with the Black Cloud game, I call the sequence of learning that the rest of this chapter describes "inform, perform, transform." As its name suggests, there are three thematic components to this approach within my classroom, and these themes had distinct activities tied to them:

- **Inform:** Students gather, analyze, and collate information in order to produce their own, original work.
- **Perform:** Students use the knowledge and information acquired through their informational inquiries to produce and perform new work that is tied to a larger critical and academic goal.
- **Transform:** Students extend their performance toward publicly shared knowledge and action and focus on directly transforming their world.

Using this model, students construct their own body of knowledge, and the teacher acts as a facilitator in this process. These three phases of play and learning took place as part of Ask Anansi, but I also believe that the model offers a clear way to sequence learning and inquiry to help shift classrooms toward a more transformative and socially engaged hub for learning. Again, relationships are foundational to the sequence. No one game (or digital device) can revolutionize classrooms, but gaming and digital contexts can propel the pedagogy already in place.

Informing: Fostering Youth Knowledge and Critical Literacies

The first phase of this sequence—informing—relies the most on traditional models of classroom instruction. In order for students to construct powerful and interactive experiences for their peers and community, they first needed to understand the rules for play and the ways they could use the tools provided to create their own work. I modeled an engaging learning process for students, asked them to share their own knowledge about a topic, and then guided them through individualized application of this larger concept. This demonstration involved producing and hiding scavenger hunt clues rather than diving into a textbook or class novel. The class collectively analyzed clues that led them to QR code badges hidden around South Central High School. As can be seen in the opening of this chapter, the first clue was easy and inside the classroom. In that anecdote, the game moved students from tepid participation to out-of-the-seat ownership of the class curriculum. Various strategies were employed by the students. The grammar of play was learned and developed during this first, easy clue. The final clue I provided was more difficult. Several elements—the vocabulary used to describe the location, the spatial thinking required, and the fact that the clue was located in a place many students had not explored before—led to a more challenging and collaborative problem-solving process.

As a game, the scavenger hunt increased student participation for the day. A student like Dante usually was not willing to join class activities. But as a result of full student participation in the game, we all were later able to participate in a conversation about Malcolm X, school facilities, and school-based authority. We reviewed the clues' language, our processes of discovery, and the materials we encountered in our search. Some students, like Minerva in the chapter's opening, were unwilling to reflect during the game, but after all of the badges were found, students debriefed about their processes of finding and the content they encountered through this search. In question-driven dialogue, students asked why they were sent to explore specific locations and what kinds of ideas these spaces reflected about the social world of SCHS. Within our class, I facilitated these conversations with questions for understanding the sociopolitical nature of schooling at SCHS. Students relied on photographs that they took with their iPods in order to share evidence and explore the questions I asked. For example, after the entire search was finished, I tried to help students see meaningful connections between the various clues they encountered by asking questions at focused on equity, like "Why do you think the principal's desk looks the way it does? Look at your pictures if you need to"; "Can you compare the principal's desk to Mr. Winter's Armadillo?"; and "What is the purpose behind the wheels of Mr. Raskol's desk?" Playing games was tied to building trust and relationships. As much as this game was designed to support student learning, it was also about strengthening the social ties of the classroom community.

To connect thematically to the idea that Anansi was hiding these clues, the remaining badges (and the ones that students subsequently created and

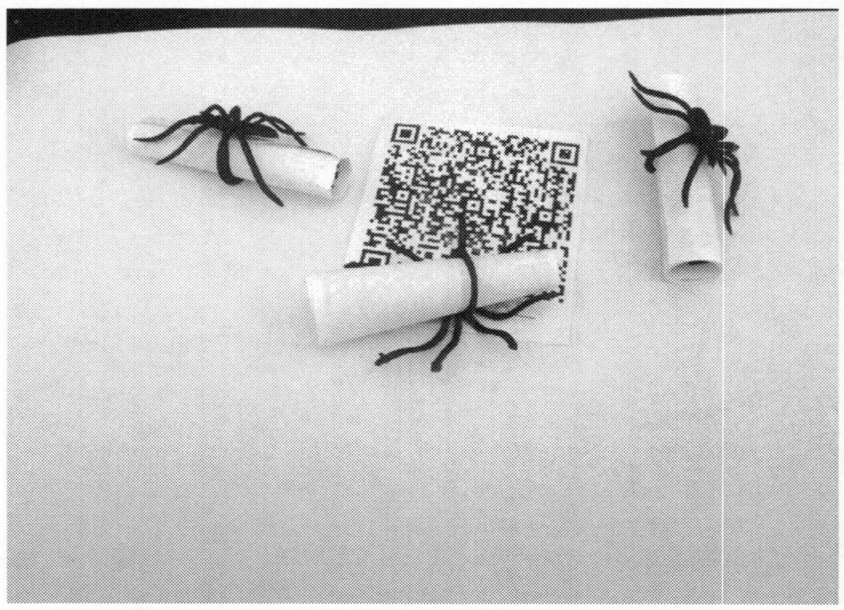

Figure 4.1
Examples of plastic spiders and QR codes that were key ingredients for the Ask Anansi scavenger hunt

hid) were wrapped scroll-like inside the ring portion of black plastic spider rings—the kind typically given out to children at Halloween.

The second clue that the class was given eventually led students to examine the portable desk that Mr. Winter used when he traveled from room to room. I asked the students why they thought I hid a clue in the bright red desk that Mr. Winter affectionately called "the armadillo." Before handing out the third clue, the class spent a few minutes discussing my intentions as the game designer for the clue:

> Elizabeth, the student who successfully located the badge in a paper tray, said, "I think you wanted us to think about the fact that Mr. Winter doesn't have his own classroom."
>
> "Maybe you wanted us to look at the big-ass desk that Winter has to move every period," Solomon added.
>
> "This school can't even afford to give its teachers their own rooms, and it will still probably fire Winter at the end of the year," Jay said, highlighting how the district's near-annual Reduction In Force process meant that Mr. Winter was regularly laid off for several months before being reemployed several weeks into the academic calendar each year.

After discovering each clue, students were required to document their findings by filling out an entry in their "Finding Log." The log prompted students to write down factual information about the clue—what it was called, who authored it, and when and where it was found. The log also asked students, "What do I think this space's story is? How does it help tell a story about the community?" For each clue in my demonstration activity, I wanted students to think about the game I had designed for them: why were we looking at the areas I sent them to?

After a second clue led students to discover a spider hidden on the principal's desk, the class was given a final clue to the "Hidden Vestibule." In an abandoned classroom space in an upstairs alcove of one of the school's buildings, students were signaled that they were in the right location by a solitary spider dangling from a string in the doorway of the unused room.

Marjane quickly grabbed the plastic spider and posed to highlight her win. Other students shuffled into the cluttered space. Katherine let out a sigh of dismay: "It's so dirty up here. Oh, my god." She said this while holding her iPod up to eye level and snapping pictures of the disarray. Holden and Elizabeth also walked into the room and began snapping photos. Dante brooded in the back of the pack, dismayed that he'd discovered only one of the four badges for the day.

Katherine and our other coresearchers internalized the process of documentation in a fluid way. After nearly a month of being asked to photograph, document, or text qualitative data about their community for Anansi, doing these activities was now a natural performative element of how the students understood their role when engaging in academic work.

Although students later pointed to the scavenger hunt—both the model described above and the version outlined later in this chapter—as a "fun" component of the class, it was designed as an intentional step forward in how students socialize as part of their learning process and interact with their larger school community. Over time, the competitiveness of hiding and finding class clues became secondary to having the language of the clues challenge traditional, dominant narratives about school spaces. Developing counternarratives was a key component to the gameplay sequence.

After practicing how to scan and create QR codes (discussed in the previous chapter), my students used a QR app to create, publish, and distribute

Figure 4.2
A student showing off the plastic spider that revealed a clue's location

counternarratives throughout the school. The production of these codes could have been framed as rote and procedural, but instead, the technical literacies were couched within the explicit purpose. By developing these QR codes as part of the class gameplay of Ask Anansi, students used applications and components common on mobile media devices as tools for authentic and critical learning. Other than some online research, using QR codes was one of the few aspects of the scavenger hunt that required digital media tools. Further, although the QR code wasn't a crucial element of the scavenger hunt (all of the elements of hiding and of seeking would still have been in place regardless of whether QR codes were used), using

Figure 4.3
Students exploring an unused classroom in an otherwise overcrowded school

technology to access and engage with the codes made the gameplay feel more clandestine than if the game had been played without this digitally produced piece.

Performing: Producing Clues and Tangling with Webs

By the time the class finished the sample scavenger hunt that I prepared for them, students were excited about creating their own scavenger hunt clues and hiding badges. Minerva and Dede formed a team and began plotting locations to hide clues. Likewise, Dante appeared thrilled about the competitive prospect of potentially finding clues before others. Instead of taking his usual seat in the back of the room, Dante moved his seat closer to the front to show me the progress he was making and to ask questions about his writing. As I looked at the class, I could see students connected to their work and to their peers, regardless of which aspects of the game they were interested in.

Students had opportunities to write clues and hide scavenger hunt badges in pairs before producing them independently. Over a one-day mini-search, students collaboratively wrote clues about areas located within the classroom as practice for writing individual clues. This activity led to an influx of clues written about the colorful closet in our classroom. Tess labeled her clue for this space "The Reckless Closet":

> Notice how a chair has to hold up the two doors for this closet. Also how there's tagging all over it and it just looks like garbage now. This area relates to the reputation of the community because our community has tagging anywhere & everywhere. Also because no one tries to take care of our community and it's like if we basically are dependent on other people to take care of it for us. Wouldn't you say the same thing?

Students were now selecting and writing their own clues, and the game Ask Anansi allowed the class to become a hub for purpose-driven production and feedback. Although I asked each student to hide a minimum of four badges and turn in four clues, nearly half of the class turned in more than four clues.

In a clue titled "Captain Green!," Dede drew her classmates' attention to the lack of plants and other vegetation on campus:

> We all know where we study but we don't seem to take a look at this Green friend of ours. After school some of us walk right next to it its real close. Notice how you see this everyday when you come in. Why is Captain Green all alone?

Dede's clue is similar to many of those written by her peers—simultaneously playful and provoking. This clue shows critical thought while it also allows students to enjoy the scavenger hunt as a game. As with the demonstration, I asked students to structure their clues so that players would look at and question the spaces they found themselves in. Students also were encouraged to write clues that provoked questions about issues of equity and the class research questions. In addition to being a part of a strategic game, the clues were the beginning of a dialogue. Ideally, when players searched, they were forced to look at and *read* the locations where each clue sent them. For example, Minerva's clue, leading to a spider taped under the building's water fountain, helped her classmates look at how the poor conditions of her school environment had been normalized. In the middle of her clue, she wrote: "We use it or have used it before but still we think it comes out nasty. Its where we walk by everyday and sometimes we don't even look at it."

Figure 4.4
"The Reckless Closet"

With three days to write and hide their clues, students in class spent much of the week busily creating and revising the language for the hints they wrote. Precision was important. Meanwhile, the class service worker, Bruce, was helping with the preparation of the clues. Because SCHS does not have enough elective courses to offer juniors and seniors, those students may choose to act as classroom aids. Bruce usually asked for permission to focus on his own schoolwork, but for this exercise, he was also in the classroom helping students understand assignments, answering questions, and preparing to disseminate dozens of clues to the students. Bruce acted as the collator of clues: as students revised clues, Bruce made copies of each clue and collated them for impending distribution.

Figure 4.5
Dede attaching a clue to "Captain Green!" on one of the few plants on the South Central High School campus

On Friday, each student received a workbook of more than eighty original clues. The first person to solve a classmate's clue and find their classmate's QR code could claim victory over that space. With a class period and a weekend to search on their own, students hunched over clues and copies of campus maps. During the class period, students were encouraged to develop strategies to find as many clues as possible, and they scanned their maps, plotting how to retrieve clues based on where they speculated they were hidden and how to do so efficiently. Two students—Elizabeth and Marjane—grouped clues by proximity to each other to find badges that were closest to each other and built a mapped journey of searching across the campus. Elizabeth started by searching for several badges that were in our building before venturing to the outside quad, the neighboring building, the main lunch area, and an area near the school's auditorium.

As the narrative of the Ask Anansi game prepared students to hide and search for each other's badges, it also moved toward increased action. This narrative culminated in a sharing of the storytelling duties in and about their community. In doing so, Anansi revealed that members of the class were recruited as "Agents of Anansi." Students were given name tags that allowed them to move around the campus on specified dates and times in order to look for appropriate hiding places and to place their badges in the correct locations. Getting these name tags for students was not an easy task. After requesting and discussing the name tags with school administrators, I also needed to contact all of the campus security officers to notify them that my students would be out of class for a day and would not be wearing the appropriate hall passes (there were still minor complaints, and one student was sent back to class). These passes allowed students to do more than just bypass security, though. By wearing identification as "Agents of Anansi," students found that their identity shifted once again. No longer were they simply doing "English work" or playing a game. Students could interpret their actions as embodied role playing, a specific identity practice.

As students shifted their roles from hiding clues to seeking them, they used different gaming strategies and approaches to dealing with the bulk of data that was presented to them. With such a large stack of clues to work through, students approached their assignment in very different ways. For example, Dante, donning his "Agent of Anansi" badge, took his clues and immediately left the room. Unaccompanied, Dante ended up finding more clues than any other class member. In contrast, Elizabeth slowly riffled through the clues, sorting them by location but also by what she felt was level of difficulty. She also looked at the gaming strategies of others, saving for later clues "that [Dante's] probably already gone found by the time I went out there."

Dante's Mobile-Mediated Journey

Before looking at the final phase of this learning sequence, I want to pause and look at the kind of student participation and growth in writing and critical dialogue that I observed emerging from my students as they played the Ask Anansi alternative reality game. This was not about appropriating youth interests or tricking students into learning. Rather, the foundations of powerful game-driven learning experiences were tied to student identities. For example, I want to share how Dante's competitive nature emerged during the Ask Anansi game.

Reading over my notes from the week when students first received their iPods, I found the following reflection:

> Dante, in three days, has gone from not participating at all to explaining assignments, leading, and arriving early to do different kinds of work. I need to figure out how this can be translated into grade improvement since he failed last quarter and what in the class is causing this radical reinvention.

Throughout the remainder of the class, Dante's work was staunchly independent and focused. Choosing to sit away from the rest of his classmates, near the back of the classroom, Dante often hollered responses to discussion prompts and became our class's natural heckler and champion. Similarly, the alternate reality game play allowed Dante to be highly competitive. In the scavenger hunt sequence of the game, he was driven to find the most hidden badges.

Dante enthusiastically participated in this project's scavenger hunt and curation activities, but here I focus on a specific interaction within the class that displayed Dante's growth as a leader and shows how schooling can be affected by participatory culture and how games and mobile devices can aid in this transformation.

But first, here is a little bit about Dante's academic journey in this class. Because Dante joined the class halfway through the year, we had known each other for two months before this study began. He did not have the same amount of time as my other students did to participate in shaping the class community. At the end of our first week of class together in January, I talked briefly with Dante as he was putting on his backpack. My intention was to check in with him, as I did with any new student in the class, to see what concerns he might have, how he was feeling, and what could be done to help him transition into the community. Typical questions for assessing how students were feeling included "How has your first week at SCHS gone? Is there anything I can do to make you feel more comfortable in this class?" Typical responses in these interactions are shrugged shoulders and monosyllabic responses, and they are one of many points at which trust was built between my students and myself. Dante, however, told me very clearly that his week went fine and that I shouldn't bother checking in on him.

"Why's that?" I asked.

"Look, I probably ain't gonna do any work." He then paused for a moment before adding, "Besides, at my old school, I made my teacher cry, so you should leave me alone. The only reason I come to school is because my moms makes me." And with that, Dante headed to the door and walked out.

When planning for my class, I spent time thinking about this interaction with Dante. It did not feel like a threat from Dante or a challenge. His matter-of-fact demeanor in sharing his previous experience felt, to me, like a statement of his expected laissez-faire time spent within my classroom space. Between his initial statement that he would not do work and his warning that he made a teacher cry, Dante punctuated his response to me with a deliberate pause, adding gravity and ensuring that I understood his power as a student.

We now fast-forward to the scavenger hunt activities described earlier. In addition to other assignments, students were writing their own scavenger hunt clues and hiding the badges (and spiders) throughout the school and surrounding neighborhood.

Dante gets to class early. I'm erasing notes from the period 2 discussion and selecting music to play. Dante runs in, pulls the door shut behind him, and scans the room to ensure that all of his classmates are outside. He tells me he has to hide his badge and spider and asks me to leave. Even though I won't be participating in the scavenger hunt, I oblige, walking out and offering greetings to students as they walk by.

Dante emerges a moment later smiling.

Twenty minutes later, after silent reading, I give students their "Daily Journey Gauge," a worksheet that allows students to self-assess their understanding or progress on a specific topic. On the board, I write the topic: "Knowing how to write clues and hide badges."

As I walk around looking at student responses, Dante races through the writing on the paper. Within two minutes, he hands his worksheet to me.

In one section of the paper, students are asked to write when they know they are successful at their task. For this section, Dante writes, "I hide clues that no one finds."

I put my finger on Dante's paper below this line and say, "Dante, when I hid clues for the class on Wednesday, do you think I intended for you to never find them?"

"No," he says while typing on his iPod.

"What do you think I wanted?" I ask him. "What was my goal?"

"You wanted us to look for the clues and figure them out on our own."

"Yes," I concede. "And why?"

"So we'd figure it out," Dante sounds irritated by the incessant questioning. "Whatchu mean?"

"I mean, why did I put the clues where I put them?"

"You wanted us to see those places and think about them?"

"Yes, exactly. So what is the goal if you are writing clues?"

"For people to struggle before finding them right away, but to find them at some point."

"Cool. So would you mind revising your response on your worksheet?"

Dante scratches out his answer, draws an arrow indicating that readers should flip his paper to the other side of the page, and writes, "I write clues about places that I think about that people struggle with but eventually get there."

Moving on to look at other students' work, I notice that Jay has a similar statement: "Write clues so no one finds it." I tell him I would like him to revise it.

Jay says, "It's alright, I got it."

"I'm concerned that you've written that it will be successful if people don't find your badges," I say. "Ask Dante why this concerns me."

Jay looks to Dante and nods, "What's this about?"

Dante, looking up from the iPod he has returned to tapping on, gets out of his seat, and stands above Jay's desk. He looks at what Jay has written, and as I walk over to help other students, Dante grabs a chair and sits next to Jay. When I return a few minutes later, Jay has not crossed out and rewritten his response, as Dante did, but instead has added a caveat to his original statement: "Write clues so no one finds it in ten minutes."

Dante has returned to his seat, his earbud playing rap music and his fingers dancing above the glowing screen.

My conversation with Dante was the type of interaction that helps guide principles of game design. Players like a challenge, but if the solution to a puzzle is not clear, the puzzle offers little reward. Rather than show this to Dante, I helped him to develop this understanding through dialogic reflection on our past experiences in the class. And despite his earlier warning to me about his probable lack of participation in my class, Dante relied on independent engagement to move forward in the class. Allowing him to hide his clue while I stepped out of the room was a sign of respect and trust. When he initially resisted my suggestion that he revise his writing and his behavior at the beginning of the class, Dante showed that although he enjoyed aspects of the class curriculum (like hiding and finding clues), he often felt disconnected from the classroom community.

Perhaps most relevant about the changes in Dante's engagement in class are the ways that the context for learning shifted. Because the Ask Anansi game expected Dante to communicate with fictional and real individuals through informal channels of communication like text messages and because the work Dante was asked to create—from QR codes to digital productions—looked different from traditional work, he was able to interact within my classroom in ways that didn't feel like school to him. In fact, by mirroring the public sphere, the learning environment that we were sustaining was a markedly different space than what typically was found within the walls of SCHS.

Despite his transformation, Dante did not do this work in traditional ways. If other people were to visit the class, they probably would have seen Dante's behavior as defiantly resistant. In notes taken during this study, I observed that Dante

- spent most of his class time on his iPod and listening to music.
- scribbled his work quickly when he finally did it so that—as he explained to me–"you'd stop pressing me."
- Acted resistant to criticism when I challenged him about work that I felt missed the mark. Typical comments included "Naw, it's good" and "It's *done*, isn't it?"
- Played music from his headphones loudly enough that it interfered with other students during silent reading in class. At one point, Elizabeth sighed as she looked from Dante to me and asked me if I was "going to do anything." Although Dante turned down his headphones in this situation, saying "My bad," he turned up the volume again later in the same period.

During the first months of my engagement with Dante, his lack of work and his aggressive response to my attempts to bring him into the class community often left me feeling defeated and ineffective as an educator. What is striking to me about the changes that he underwent in class are not that they are due to the implementation of technology as a tool for learning in class but that these changes were *revealed* when technology was implemented as a tool for learning in class. Long before he received a school-issued iPod, Dante was recreationally engaging with his phone in my class. However, any academic merit that might have come from this work was not something I could see. For instance, prior to handing out the iPods, I found Dante tapping at a dizzying pace on his phone during the class's silent reading time. When I asked him what he was doing (and assuming he was playing a game), Dante said he was writing down rhymes. Expressing interest in his extracurricular writing, Dante offered to text me some of his rhymes, which he continued to do throughout the year. Students like Dante frequently and without coercion engaged in the literacy practices that English teachers promote, but the traditional classroom typically renders Dante's efforts invisible. As my own initial assumptions reveal, their efforts often are not only invisible but also seen as hostile to the learning environment.

To reiterate an earlier point, from a traditional perspective, Dante's academic work did not shift significantly. His formal writing of texts using Standard English remained rushed and inconsistently completed. However, as much as Dante deliberately signaled that he was not a strong academic student,[19] he emerged as a leader within the class. Because Dante took ownership over the gamelike elements of Ask Anansi, he was able to apply his own interests to how he guided his peers. Additionally, instances of his engagement that were previously invisible in the traditional classroom space—when he played games or texted his rhymes to me—allowed me to see Dante's role in the classroom differently. When the class discussed texts, QR codes, and game strategy, Dante appeared more comfortable having his work critiqued and challenged by both peers and myself than he did when we discussed traditional topics. The exchange in the above vignette, where I challenged Dante's writing about the definition of success within the scavenger hunt activity, would not have been possible without the shifts in engagement he experienced in the previous weeks. The exchange that followed, in which Dante aided his classmate in understanding, also probably would not have occurred. It is not that technology made Dante a better student. He didn't transform as a student the moment the Apple logo appeared as the iPod was booting up. Instead, the device and the shifts in my curriculum around it allowed Dante's already existing learning strategies and expertise to emerge and for Dante to contribute positively in the classroom.

In reflecting on Dante's engagement in my class over the course of the year (at least the part of the year he was enrolled), I am reminded of the *comfort* that digital devices provide in society today. In any public space today—a mall, a restaurant, the student union of a university—you will see young people and adults enraptured by their mobile devices. I distinctly remember the moment of cognitive dissonance I had when walking through a mall in Los Angeles in the years just prior to mobile phone ubiquity and seeing a couple strolling along, holding hands, and both talking on their phones, presumably to two different people. At the time, I experienced a feeling of cold alienation in response to their behavior, and yet today this is a common sight, and I myself am often guilty of this kind of disengagement from the physical world when *enchanted* by my phone.[20] Waiting for an appointment, killing time on a plane, or sitting in a bar waiting for a friend, we are never uncomfortably alone when a mobile device is at hand. A few swipes through a social network, a brief check-in with a friend, and a casual attempt at a high score on the latest freemium game: we are never alone when we have our phones. I am not saying this is a good thing, but it is how many individuals have shifted their participation in public life.

Despite having some concerns throughout the study, I found that my students felt comfortable using the digital tools (such as audiobooks and text messages) that I provided in the classroom. Many students elected to squint at an e-book of *Little Brother* rather than simply read a traditional paper version of the book. By building on the comfort that young have with their devices, I was able to ensure that the learning practices tied to these devices were no longer rendered invisible within my classroom. Considering the ways that Dante emerged as a leader because of the visibility of his learning emphasizes how wireless tools can help reinforce trust between students and teachers. Although not a panacea, within the context of my classroom these digital tools mediated critical conversations that yielded leadership, agency, and academic engagement.

Dante's engagement via a game reinvigorated the ways that traditional leadership can be identified and illustrates how his youth expertise reshaped classroom interactions. This transformation would not have emerged within traditional classroom pedagogy. Unlike Minerva's empowering learning driven by trust, Dante's engagement was furthered by technology, games, and competition—attributes of my classroom that would not have been present without Ask Anansi. Dante taught me that all too frequently students in our schools are disengaged because we remain too entrenched in practices of the past. Like my students in the final phase of this study, it is time to take a hard look at the schools we are sending students to today and consider what aspects must be transformed.

Transforming: Fair Play and Webs of Critical Action

Students were frustrated with the pacing that I initially envisioned for the class. Whereas I had hoped to have meaningful dialogue at the sites of each scavenger hunt location, many students were antsy to find the next clue. Rather than force this issue, I adjusted. Instead of debriefing after each clue, we collectively returned to the classroom and reflected on all of the locations. After students completed the scavenger hunt, we discussed what students encountered on their journey, who the class considered "the winner" of the hunt, and what our next steps would be. Because most—but not all—games are competitive,[21] students were intent on comparing the results of their search and determining a winner. I wasn't prepared for this.

Looking back, it seems inevitable that of course students would want to identify a winner of a competitive game. I simply hadn't anticipated this. It was another shortsighted aspect of the design and one that I attempted to—albeit inelegantly—address. Because I did not anticipate having to select a "winner," I attempted to shift the class discussion to focus on how the game could be taught to others. We engaged in a discussion of what fairness means. No students ever raised concerns that the game was unfair. I used this framework to encourage students to think about what each student "won" through the experience and what the value of that achievement was. Moving discussions of victory from the typical zero-sum models of board or video games helped elevate the classroom community discussions further. Additionally, in a poststudy interview, students described fair play in the game by saying that students independently searched for badges and relied on no other individuals while searching. I pointed out parallels between gameplay and school structures—that fair gameplay and equitable gameplay essentially establish similar patterns of equality.

For students, fair also meant making a "reliable" clue. As Jay explained, "Some clues were just like finding a piece of grass on the lawn. It was impossible." Some clues were so narrow that following them was more tedious than stimulating. Even clues that were too detailed still offered nuanced critiques of the spaces they directed students. As Holden said when discussing the single blade of glass example, "I wasn't going to search every piece of grass on the field, and I could see the point of the clue—the field is dirty and looks all dangerous."

As we continued to discuss the game, Elizabeth opened up a discussion of what made playing successful. For her, winning this game was about both designing and playing. She felt engaged when writing challenging and enlightening clues and searching "with all your energy" for the clues that could be found.

Looking over the critiques that students created of their school space (including several nearby off-campus locations like a fast-food chain next to the school), the final step was for students to invert the game. Taking ownership over the spaces and narratives the students created, the class became a space of museum-like curation. Selecting two of their locations, I asked students to write very short descriptions of what they encountered, along with an informative or intriguing title. The assignment also asked students to conclude their descriptions by posing a question to the individuals who might read them. After these cards were written and revised, students posted them prominently in the spaces where they originally hid clues. The hidden badges of the previous part of the game morphed into prominent public displays of knowledge and dialogue. Minerva's clue originally was hidden under the dusty water fountain outside our classroom, for instance, and she later wrote it on a placard, which she taped onto the water fountain, unavoidably within the line of sight of anyone using the fountain.

Student thinking and voices were now on display publicly throughout the school. As educators and policymakers explore ways to weave participatory and mobile media practices into classroom practice, this activity suggests the value of mobile media to empower not only the individual learner but the community at large through easily replicable models of instruction. Incorporating mobile devices and participatory learning here did not require fancy gadgetry. They facilitated learning throughout, but the shifts in classroom experiences did not rely on technology. Although students produced nondigital products like the note cards in this sequence, they were empowered to do so through a wireless pedagogy. Connecting to their

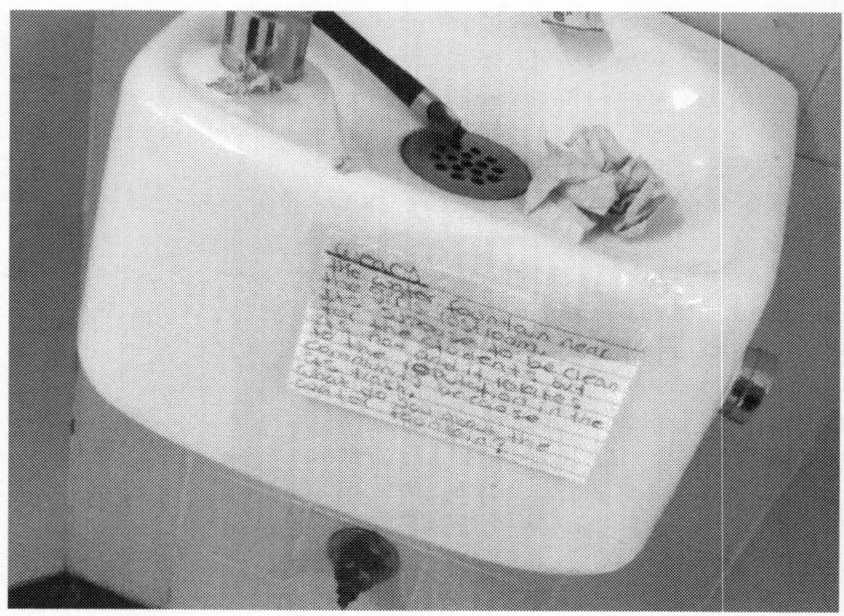

Figure 4.6
Student-written placard curating a school space and highlighting school inequities

community and documenting this process moved learning beyond writing on paper.

Further, the transition from making this a "closed" game that was available only to the students in my class to making it an open dialogue with the community is—to me—the most important element of this curricular endeavor. Like graffiti or billboard advertisements, curating the space and reconfiguring how people see their environment is invasive. The note cards that students wrote could be removed—nothing permanent was added to the school's spaces. They confronted the public with counternarratives about how each space was used, valued, and interpreted. This was an extension of the entertainment that students felt while playing the game. In an interview after the game's conclusion, Solomon said, "The hiding and the finding and the writing those little cards, that was dope." Hearing Solomon's affirmation, Katherine added, "It wasn't just that it was cool though to write the cards, it felt like, I don't know, like the way to explain this work to our friends."

In writing placards and physically curating their experiences of their space, students did not simply "read the word and the world," as Paulo Freire and Donaldo Macedo have described sociocultural literacy practices.[22] This process also was one of *writing* the world. The students' peers encountered this work, and Katherine's statement reflects an understanding of the role that peer networks play in building critical thought. Katherine highlights how the curation phase of Ask Anansi can lead to an ongoing integration of students as they share and engage with their community.

Even with all of the tangible production of work, students also engaged with the digital world. As part of their class assignments, they edited the official *Wikipedia* page for South Central High School using the conventions of the website. *Wikipedia* has played a somewhat controversial role in schools. Teachers have frequently challenged the validity of its sources and content, and students have often been critiqued for using the site for its presumably shoddy content. However, *Wikipedia*'s base of contributors has grown significantly since its inception in 2001, and its value to the educational community as a place for both consumption and production of knowledge is significant. danah boyd's analysis of the role of *Wikipedia* in schools and research for teens has pointed to this shift.[23] Further, the openness of *Wikipedia* and the robust dialogue that happens in the comments and history pages for many entries emphasize the nonobjective nature of how knowledge is shared and produced.

When I demonstrated how *Wikipedia* works and how edits are almost always automatically undone if they are anonymous and lack properly formatted citations, the class explored the content about their school on its *Wikipedia* page and the writing and formatting conventions of the site. Students captured qualitative data regarding student opinions about the food, emphasized the school's dropout rate, and noted the small custodial staff for the large school:

> As of 2010, the dropout rate at [South Central High School] was 68%.[9]
>
> With more than 90% of students qualifying for free or reduced lunch provided by the Los Angeles Unified School District,[10] some students have noted that sometimes meals are not heated properly.

Students included details on the *Wikipedia* page that were just as factually sound as the existing information about the school's history and alumni but that allowed them to voice what they felt was necessary, important, and relevant to the people who might access the page. Footnotes 9 and

10 in the excerpt above pointed readers to the 2010 UCLA Educational Opportunity Report.[24] Holden and Jay, the students who felt it was necessary to add the school's dropout rate as additional information to the page, struggled with the website's platform. Each time they clicked "Save page" on the SCHS screen after making changes, their changes were removed seemingly instantaneously. Not until the students copied the formatting code for previous citations and replaced the text with their own linked source did the *Wikipedia* platform accept their edits. More than a year after the students made their changes, those additions to the *Wikipedia* page were still intact.

Conclusion: Toward an Ethos of Classroom Hacking

The Ask Anansi game is largely about taking pieces of academic and civic learning and reimagining them as points on which to connect play, imagination, and a healthy dose of story. It is a hack of traditional pedagogy. Classrooms can be hacked by playing games and embedding a narrative arc into the activities that students complete. Although the word *hacking* originally was used to describe harmless tinkering with technological devices, it now often is used to describe a malicious act. The mainstream media publicize hacking scandals, and the general sense is that hacking is done to individuals by strangers. These connotations of the word make it seem like a dangerous learning practice that that we should warn youth away from. At the same time, ethnographic accounts and research analysis of the efforts of hackers demonstrate that only small portions of this community act maliciously.[25] In these studies, hacking is the productive and collaborative effort of programmers as they respond to the needs of users. For instance, in looking at the history of open-source software development that has led to popular software such as Mozilla's Web browser, Firefox, Eric Raymond describes the hacker communities as operating as a "bazaar"—where people trade ideas, code, and innovations to help the general populace.[26] In contrast, Raymond delineates traditional nonhacker production (such as companies like Microsoft) as operating more as a "cathedral"—where teams of designers spend a digital eternity creating a product that could be outdated as soon as it is completed. In this cathedral and bazaar model, hackers promote productivity and spur innovation. The model illustrates much potential for possibilities within an academic context.

A hacker ethos points to how literacies are expanding. When students hack online spaces, they copy, remix, and recontextualize information.[27] The "inform, perform, transform" model is one of reappropriating and hacking the spaces and possibilities of schools. By its nature, hacking requires a synthesis of an original author's content, intentions, and form. Students need to be able to understand the nuanced genre within which they are performing, producing, and hacking. Hacking is almost always productive: something is *made* through one's hacking efforts. Like a five-paragraph essay, a research report, or a video composition, a hacked site is a newly created production that depends on the works and contributions of prior authors.

Hacking engenders an individual within a community of practice.[28] Studies of hacking communities of both adults and young people demonstrate how individuals in such hacker communities freely share the expertise of all of the participants;[29] myriad voices are valued in such an "affinity space."[30] As noted previously, the hacker bazaar becomes a space for exchange.

As they hacked existing dominant narratives about their communities and reappropriated space through their curation as part of the scavenger hunt, students playing Ask Anansi shifted how learning occurred in schools, where it occurred, and for what purpose. At the same time, this process was not one of purely hacking old forms of learning. Traditional aspects of academic engagement can be seen, too. The academic process of "properly" editing the school's *Wikipedia* page represents how traditional learning can be harnessed for critical outcomes.

Although students used mobile devices to support the hacking ethos of the classroom, I also wanted this exercise to show that twenty-first-century "wireless" learning can take place without an overreliance on complicated digital tools. In fact, as I reflect on the role that gameplay can play in fostering points of participation in a class, I think that the digital aspects of games often get in the way. Reflecting on the role played by the kitchen or dining room table as a place for families to engage, noted farm-to-table chef Alice Waters said that

> The table is a civilizing place. It's where a group comes and they hear points of view, they learn about courtesy and kindness, they learn about what it is to live in a community—live in a family first, but live in a bigger community. That's where it comes from, don't you think?[31]

Waters reminds educators that what brings people together are the spaces and cultural rituals that tie us together. Although play is an innate part of human development, digital tools have largely divorced it from its social roots. Technology augmented student talk in the Ask Anansi game by staying out of the way. Students regularly engaged with mobile media throughout this project—documenting evidence, looking up *Wikipedia*-related research, and writing discursive notes as both game designers and game players—but technology was never the spotlight of the scavenger hunt sequence of Ask Anansi. Rather, it was a game about dialogue and collaboration.

Getting students to move more freely beyond the walls of traditional classrooms required recognizing the types of practices that students engage in when they are learning at home, with friends, and in extracurricular spaces. It required adapting the interests and activities of my students for formal, critical engagement. Similarly, even when the pedagogical design of this game-based activity meant students were not using their iPods during many parts of this work, the project was designed to replicate online participatory media shifts. The tools of critical digital practice are insignificant. Students need in-school practices that expose them to the ways that engagement with media and with society have shifted.

5 Can You Hear Me Now? Leadership and Literacies in Today's Classrooms

It was the second week of June, and the frenzied energy in my classroom matched the substantial workload the class had looming before it. In addition to completing their curation of local spaces through the Ask Anansi alternate reality game, students were doing research related to editing the South Central High School *Wikipedia* page, writing autobiographical children's stories, and revising research reports related to their topics of inquiry. Earlier in the school year, these students had seemed unmotivated to engage in "school work," but now they were excited about writing.

When I asked if they wanted to share their work with other classes, schools, and teachers, Katherine interrupted me. She complained that these class activities were fun but never actually lead to changing anything: "We've done work like this in Mr. Major's class, too. It's fun and all, but it won't change anything."

"They don't listen," Solomon said, agreeing with his classmate. "It doesn't ever matter."

As the students spoke, I recognized their frustration. Annually, students in my classes engaged in critical, inquiry-driven research expecting to see the large paradigm-shifting changes that are reflected in Hollywood "teacher movies" and in highlight reels of protests on the nightly news. Popular media makes youth-led change look easy. The realities of slow-moving change are out of sync with student expectations. And considering the fractured nature of adult political discourse in the United States right now, the civic work of educators must include questioning and challenging cultural perceptions of young people's capacity to propel social change.

Civic education can occur in every classroom students enter. If your schooling experience was anything like my own, civics was relegated to a single-semester class for high school seniors and was largely a place where I learned about the three branches of U.S. government and my responsibilities as a citizen to vote and enlist in the Selective Service System. That's too little and too late. Civics education can be much more in today's participatory society. Civics should be an underlying purpose of how every class is taught in schools at every grade. Civics and the lessons of participation and activism should begin in kindergarten and should increase in complexity as students advance through the school system.

What is "growth" when we talk about what happens in classrooms? Who grows? And in what ways? In an era of education that is driven by high-stakes tests and the pacing plans, teacher evaluation models, and a culture of fear that comes along with them, the pressure for students to excel has never been higher. But excel at what? When educational policies push schools to be better, we must consider the endgame for students: what are we preparing them for, and how are we guiding them to untangle the civic knots that they are inheriting from us? A common refrain in the remainder of this book is that we need to rethink the *purpose* of school. As we focus on improving schools, we also must focus on getting students better prepared for guiding and leading in society, and to do so, the civic purposes of all classrooms must be considered in terms of how schools grow and change today.

Considering the challenges of civic education in an era of high-stakes testing, I return to my students' frustrations. How do we teach lessons of critically engaging civic education when students like Katherine and Solomon feel like "it doesn't even matter"? What does it mean to "do something"? As much as educators may engage kids in critical discussions, research-driven inquiry, and the seeking of solutions to local problems, translating this work into meaningful change is often unlikely. For students who often struggle to be heard within schools, this is frustrating. How do teachers teach in ways that lead to actionable outcomes for young people? Is such a goal feasible within public schools?

Throughout the time that students played Ask Anansi and used their school-issued iPods in our class, my research examined how students might be changing their sense of self, enacting agency, and producing texts that reflected civic principles. Seeking paradigm-shifting changes in students over just a few months is an overly optimistic activity, but several trends in student behavior, engagement, and leadership emerged during this study.

Playing Ask Anansi and engaging with iPods continued to support and strengthen academic learning in my classroom. Students produced, read, and analyzed increasingly complex texts throughout the game. But the more interesting shifts within my classroom were the aspirational and social changes, so in this chapter, I focus on the civic learning that took place in my classroom. I highlight the specific moments that were transformational for students and their ability to act, produce, and critique schooling and neighborhood challenges. Diving into what counts as growth, learning, and civic identity throughout Ask Anansi and our inquiry into using iPods, I discuss models of what civic education can mean in today's classrooms. These brief moments within the classroom signal how youth agency and civic education can be reframed for other classrooms beyond my own.

A significant part of a civic education is the way individuals read, communicate, and socialize. As an English educator, I believe that literacy practices are tied to civic learning. Although my research bridges two different research disciplines—civic education and sociocultural literacies—this chapter looks at expansive moments related to both areas as a means of understanding this notion of classroom "growth" today.

Why Classroom Civic Engagement #Matters

Civics lessons are about guiding individuals to act within society and their local community, so what lessons about participation and voice are being learned in our classrooms every single day? The purpose of civic education is to help young people see how the skills they learn in school can support their participation in the "real" world—the streets that students physically traverse daily and the digital world where they interact online. The role of activism, dialogue, and civic participation in online spaces (like video game communities, Facebook, and Twitter) should remind us all that the "local" world is no longer bound by zip codes.[1]

In considering the ways that schools can guide young people to participate and make positive social changes in their communities, what matters varies in scale. Many actions—such as collaborating in a video game virtual world, guiding activism in both online and offline contexts using hashtags like #BlackLivesMatter, and organizing community members around local issues— are all *real* civic activities. The lessons learned in these spaces extend youths' possibilities and identities. This fluid understanding of civic identity and education does not veer far from how philosopher John Dewey framed the purpose of "progressive" education nearly a century ago.[2] Even though Dewey was writing long before the contemporary digital version of a networked society,[3] his basic premise still holds up today. Dewey regularly addressed the dearth of engagement between "school and society."[4] He questioned how school-based learning affects the public sphere.[5] The ongoing tension is that the closed-off nature of schools undercuts the ways that youth can shape this public space.

Looking at Dewey's scholarship on schools' disengagement with a public sphere, we can see that the civic challenges of education have plagued schools for over a century. Even more distressingly, these challenges are getting worse. Particularly for youth of color in U.S. schools, researchers have highlighted a "civic opportunity gap" that cleaves the kinds of civic learning that students received based on socioeconomic contexts.[6]

Considering these challenges in light of the demographics and academic needs of students at SCHS, I wanted to develop Ask Anansi as an experience that would replicate "real-world" contexts of engagement. I also didn't want this experience to rely solely on technology. Digital learning should permeate the learning experience but not drive it. The digital terrain that young people traverse today can be a potential "civic web."[7] In *The Civic Web: Young People, the Internet, and Civic Participation*, Shakuntala Banaji and David Buckingham outline commonly held binary perspectives on new technologies:

> either technology will liberate us or it will enslave us; either it will expand our potential or it will reduce us; either it will revitalize our social and cultural life or it will take us all to hell.[8]

Although binaries can be useful for galvanizing conversations and scholarship, they do little to engage students in the mental, physical, and virtual spaces they inhabit. In fact, the divide between school and society that Dewey warned educators about now includes digital spaces.

Recent research has looked at the civic learning potential of online environments, including video games.[9] These research spaces are contested, too. For every study that espouses the value of leveraging resources like Twitter for civic engagement and organizing, there are claims that such engagement is part of the erosion of social engagement today. For some, the acts of sharing, "liking," and retweeting civic content amount to little more than "slacktivism."[10]

Rather than cheerleading on one side of this continual debate, I am concerned with what we *do* to meet the needs of our students who live in these myriad, diverse spaces. The comingling of civic participation and peer socialization in online spaces (whether good or bad) is broadening the scope of where and how my students learn. Rather than see my school as a monolithic space affecting youth's lives, I have been working to build off existing strategies for civic engagement in contemporary culture[11] and consider how classroom instruction can foster youth expertise and engagement. This work fits into a larger model of critical research that is being taken up by educational researchers globally. In this sense, a powerful civic learning model in schools today is fluid and takes seriously the "and" in Dewey's call for connecting "schools *and* society."

One place where I've seen a divide in school-based and societal learning is how dialogue and learning take place in online spaces. Over the past several years, my Twitter feed has been filled with the hashtag #BlackLivesMatter. The viral hashtag was picked up in the months following the police shooting of Michael Brown in 2014 in Ferguson, Missouri. The city's name also became a hashtag that spread rapidly, signifying rage and spearheading the efforts of grassroots activists across the globe who were unified under the networked tools of digital social media.

As I continue to watch and participate in the conversation within the hashtag and in local civic actions related to it, I am reminded of Claudia Ruitenberg's definition of *political anger*:

> the anger or indignation one feels when decisions are made and actions are taken that violate the interpretation and implementation of the ethico-political values of equality and liberty that, one believes, would support a just society.[12]

Recalling my students' frustrations that their perspectives, their work within my classroom, and their concerns don't "ever matter," I am aware that when classroom instances of learning and engagement—even when my students wrote a children's book (Minerva) and designed game activities (Dante)—are closed off from the rest of society, little change results. However, a shift from student work that doesn't matter to work that does #matter can broaden the ecology of learning and the spaces in which youth engage. From my own learning, I recognize that many of the ways I engage in the civic world of education today are mediated by the digital tools I regularly use.

This past decade has seen entertainment industries individually and collectively attempt to grapple with the changes brought about by the pervasiveness of Internet use. Nearly two decades ago, file-sharing programs like Napster financially challenged the music industry and changed how people consumed, distributed, and interacted with music. Although book publishers, video game producers, and video content creators have innovated to maintain relevancy in an era of digital ubiquity, teachers and teacher educator programs have largely ignored the changing digital landscape our students traverse throughout their day (and often well into the night playing games like Overwatch and World of Warcraft, being on Facebook, and text messaging with friends).

Ultimately, the current policies in place and traditional pedagogies used by many teachers at SCHS limit the civic imagination of young people and their avenues for engaging in public life. Instead, the traditional power structures in urban schools, their disciplinary policies, and the school-to-prison pipeline essentially define very limited ways for youth to see themselves actively participating in society.

The lessons students thus learn reflect limited and—for some of my students—"boring" pathways toward engagement. The digital on-ramps toward civic participation are cordoned off within the school's present organization. But civics #matter in today's digitally mediated world.

A Note on Substitute Teachers

In considering the real-world implications of what happens in classrooms, I want to reflect on the experiences my students had with substitute teachers during this study. Every teacher will need a substitute at some point during the school year, and I am not the exception. During the quarter that students were using iPods in my class, I was absent twice, which was challenging for students who wanted to maintain a balanced learning environment. Their comments led me to wonder about how mobile media may help stabilize an otherwise inconsistent learning experience.

At the beginning of the study, I requested a substitute for two days when I had to be out of town. This was shortly before I distributed iPods to my students. Teacher absence is almost always disruptive, and over my years as a teacher, I've developed strategies to mitigate the disruption that occurs. I anticipated that the sub would be able to guide students through the logistics of the course in the days just prior to handing out iPods. Rather than writing a prescriptive document that guided only the sub's actions, I wrote lesson plans as a letter addressed to the students and asked the substitute teacher to give copies of the letter to students so that they could follow its directions. In this letter, I asked one student to send me a text message telling me which students were present and which were absent so I could adjust my lesson plans for the following week accordingly. By having students report attendance numbers to me, I was trying to build trust with them. However, instead of receiving a student text, I instead received an image of the attendance roster and the following message from the substitute teacher:

> They r very lazy today, tried to help them out with answer chart n 80% dint want to participate. Now I have them working on their own. ... Take care and sorry to bother you.

Here was another instance of an adult using a mobile device to control students and extend the purview of adult authority. Rather than allowing students to use their voices, the sub sent me his perspective on what the class was doing and their attitude toward the assigned work. He detailed problems he noticed in the class and reported attendance numbers to me instead of having the students do it as I requested. Technology functioned as a tool for surveillance and for adults to talk *about* students rather than *with* them.

The second time I had a sub was one month into our study using iPods. After returning to my classroom, I was warned by a teacher who shares our classroom space that my students had had a difficult time with their substitute. My colleague, Mr. Winter, said that the class seemed to be not as much disruptive as unwilling to accept derogatory remarks from the substitute. He told me that two students—Solomon and Dante–"acted transformatively" and removed themselves from the class to avoid an escalating situation. I was curious to see what my students would say about their experience.

As the third-period students shuffled into my classroom, nearly every student expressed how much they did not like the substitute and how "mean" she was. After silent reading, I asked students to write reflections about their experiences, and we then engaged in a Socratic dialogue.

Listening to but not yet responding to student complaints about the sub, I asked the class to help me capture the data that they were sharing. Too often, the expertise of students within schools is disregarded. Most adults have experienced life as high school students, but that experience is in the past. Students today can give us sharp insights into their schools. And so I asked the class to help me use our mobile devices to document our conversation using a mindmapping application Dante had previously helped install. The mindmap allowed students to connect the main points of the discussion visually and to organize, edit, and move branches of their map to help them interpret the ideas their classmates shared.

As students opened their applications, a flood of suggestions for what to put on the map were shared loudly. Katherine began the conversation. She explained, "She [the substitute] was talking about how we so ghetto and we don't know anything and that everything's better somewhere else."

Almost immediately, Minerva added to the statement by noting that the substitute "was comparing us to somebody else—like our community and shit. … She said that supposedly we were being rude and that that represents our parents."

Elizabeth said that the substitute told her, "You acting stupid," before adding, "'You [and Zatarra] are the only two African Americans in this class,' and something about representing yourselves, 'Why you acting like that?'"

During this exchange, many students spoke at once. Some added contributions to what the substitute did or said ("She asked if you from Africa or something like that"). Others disparaged the substitute's appearance ("She's the one looking poor in her old skirt"). The conversation was passionate. The treatment by the substitute created strong feelings in my students, and I could see they were united in conveying frustration, clarifying the language and tenor of the class, and generally speaking about how they were disrespected.

I felt frustrated for my students and heard the increasing volume around me as my students vented about yet another adult telling them why they were wrong, why they didn't measure up to society's expectations, and how they were a disappointment. I wanted the students to be able to do more than complain and "transformatively" leave the classroom. As the class became filled with the excited chatter of indignant students, I finally spoke. I was angry at how my students were treated and also knew how they can come off when they are upset. I did not doubt

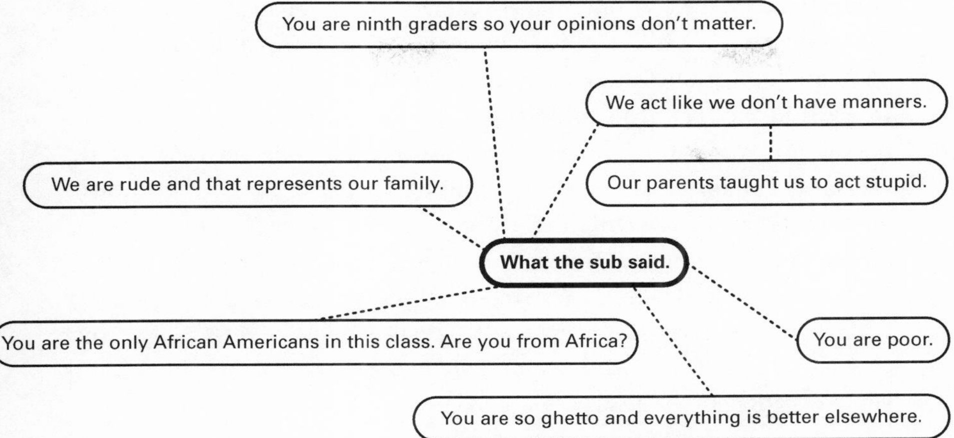

Figure 5.1
Minerva's mindmap of the class discussion about a substitute teacher

that some of the language they used with the substitute could have been disrespectful:

Antero: Here's the slippery slope. We can take our anger, and we can criticize her appearance and the things she said, or we can look at this more critically, right? I just got a bunch of data from you, right? It sounds like we received an injustice in this class on Friday. For some of you, this might have been a sign that you didn't want to continue going to school with these kinds of conditions.

Precious: I don't.

Antero: Do we just sit there? We're just ninth graders, and that's okay that that happened?

Minerva: Hell, no.

The students then decided it would be best if they organized their complaints and shared them with the school administrator who works with substitutes at the school. Minerva offered the mindmap that she created during the class discussion as the basis for the list that would be delivered to the administrative team.

Although her mindmap does not capture the robust nature of the class conversation, Minerva relied on a tool to capture the main components of my students' concerns about the substitute teacher. The mindmap provided an itemized perspective of critiques that the students had about the substitute and served as a record of her comments. It also allowed the class to *slow down* and reflect on and take inventory of the injuries they felt they had received in the class. This mindmap could have been generated without a digital tool. Minerva's predilection for using her mobile device didn't significantly enhance the quality or intellectual depth of the work. However, it ensured that she was closely engaged in sharing in the class conversation and in documenting it digitally. This digital tool is not critical by nature, but when leveraged within a classroom to build student empowerment and to critique the power structures that students experience (in this case, that of a substitute teacher), this mindmapping application contributed to the larger experience of critical learning in classrooms. The *context* of how technology was used was transformative. Technology involved students who otherwise wouldn't be interested or engaged. Tools in and of themselves do not fix classroom conditions or learning experiences.

The students' complaints were brought to an administrator, who said he would take them seriously. The class did not have another sub for the remainder of the year. The tangible changes that occurred with regards to this sub's placement at the school were not seen by the students. But this experience showed teenagers of color in South Central Los Angeles that they needed to be able to *act* on the academic lessons they learned and the lessons of injustice that they engaged with socially.

With both of these substitute teachers, mobile devices were not a major concern facing students. Even when I attempted to use mobile media as a means of establishing continuity and communicating with students despite my physical absence, my efforts were thwarted by the subs. The daily microaggressions that students experience in schools (as shared earlier in this book) are not necessarily mitigated by technology. Although my students and I were able to use their experiences as data for the class inquiry into school equity and to produce counternarratives that reframed the day-to-day lived experiences of students at South Central High School, students are still dehumanized and experience frustration that persists beyond conversations about technological innovations. The incidents also illustrate the need for a disruption of schooling practices that leave students feeling hurt, angry, and aware of their marginalized place within America's schooling system.

In Practice: A Need for Parent Dialogue

Substitutes weren't the only hurdles to creating a wireless classroom learning environment that could mimic civic contexts within society. During the week that I handed out the iPods to students, I asked them to bring back signed forms from parents and guardians acknowledging the use of iPods in the class. The purpose of these forms was twofold:

1. I wanted parents and their children to accept responsibility for the use of these devices while at school. Although I did not (and did not intend to) ask parents and students to pay for lost or damaged iPods, I hoped that by signing the form, students and parents would handle the devices carefully.
2. I also wanted parents to be aware of the pedagogical shift within the classroom. Just as in-class engagement could look distinctly different when using mobile media devices, homework also could take on a different form than what parents were used to seeing.

Because working with participatory media included texting, producing media, and communicating with peers, homework was about engaging in texts and writing individually and also about networking and strengthening classroom ties. Such work would likely reflect the informal learning practices of connected learning.[13] As a result, the same somewhat discomforting shifts in classroom activities and homework could concern some parents. I sent the letter home to both inform and invite dialogue with me about the changes in work habits anticipated within our classroom.

A day after handing out the sheet informing parents of changes within the classroom, Solomon came up to me before class and handed me his iPod. He said his mother didn't want him to use it: "She says I'm not responsible enough to be using it, I guess."

To ensure that Solomon's mother and I understood each other correctly, I called her later in the afternoon. (We had spoken several times throughout the year about Solomon's academic needs.) As I described the work that I hoped Solomon would be doing while using the mobile device, his mother patiently listened to me and suggested I let Solomon use the iPod during school hours and to hold onto it for him before he went home, ensuring that it wouldn't be lost, stolen, or damaged.

Up to this point, Solomon and his classmates had experienced school policies that communicated that mobile devices were a problem. Despite messaging from district, state, and national educational agencies that increased use of modern technology will positively affect student learning, the use of mobile devices in schools is restricted. In separate research I conducted with Thomas Philip, we found that students who used smartphones provided by their teachers usually left them in their lockers and did not actually use them.[14] These phones were stripped of any social or "fun" features and also were accompanied by strict warnings that any repairs would have to be paid for out of the pockets of the working-class students and their families. The warnings and lack of features drained these phones of the social value that they otherwise would have had. Another reason that schools like South Central High School restricted mobile media use was that they could not control or filter content that students encountered during school hours.

Although I understood Solomon's mother's concerns, they were more evidence that with students, two phones are never better than one. The meaning and value of personalized, mobile learning is lost when we separate academic and social contexts of engagement.

Ultimately, Solomon used his iPod during the school day and returned it to me before going home each afternoon, finding time throughout the day to listen to the class audiobook, text me written homework assignments, and add new songs, games, and images to it.

My interaction with Solomon's mother reminds us that even within a classroom community where I attempted to enrich learning with the possibilities of a more democratic learning environment through mobile media, the conversation must still extend beyond the walls of the classroom.

Not talking earlier and more in-depth with parents was yet another significant oversight that I made during the design of the research. It was an example of the kinds of missteps I made in not anticipating how all participants would understand the new contexts of community and learning I was attempting to craft. This imperfect anticipation also extended to gauging the expectations of my students and is most clearly illustrated by a process of (literally) rebuilding my classroom on the first day of school.

Making Change and Unplugging Learning

It was the first day of the school year. I had had an epiphany the previous week and shared it with another teacher: "Space, man, is the main problem in our school. We need to transform it but not without students."

I was sure I'd figured this whole teaching thing out. If the ways we design classrooms is often about control, then maybe I could design my classroom so that it reflected more democratic ideals for my students. Better yet, maybe the process of designing the classroom could be a democratic act in and of itself.

I'd gotten to school hours early and cleared everything from the walls of my classroom and stacked every desk and chair in a corner (which in retrospect, may not have complied with the school's safety code). I had convinced the custodian that I didn't need a teacher's desk this year. The only other items in the room were one file cabinet that belonged to the night school instructor and a wall-length whiteboard. It was perfect.

My new class of ninth graders shuffled into class collectively as the 7:26 a.m. bell rang. There were fourteen of them (although this number grew over the next few days). They were quiet. This was not only their first day of high school but literally the very first moments of it. And even if they didn't seem excited, I was thrilled for them. As the students shuffled in, they did a double take as they looked around the room. They probably were wondering if they were in the right place. There weren't any chairs or desks or any of that other stuff that makes a classroom a classroom, and the man in the front of the room looked way too excited to be there.

Giving them a minute to recheck the room number with the number on the sheet of paper that assigned them to my classroom and to let the fact that this was indeed the right room settle in, I played my ace and began the class:

"Hi there. I'm your homeroom teacher, Mr. Garcia. You probably noticed there aren't any chairs for you to sit in."

The students stood there, looking unimpressed.

I continued: "Unlike most other classrooms where the teacher decides where you sit and what your classroom time looks and feels like, our classroom is different. In this class, you get to decide how the room should be set up"—I inserted a dramatic pause—"starting now. You may not know all of your classmates, but you are going to need to work together right now and design this classroom to reflect what you want it to mean for you. Ready? Go!"

I threw my hands up in the air as if signaling we were about to start an epic race. The room was silent as the students stood looking at one another or the ground for longer than I wished. My capacity for outwaiting students was tested that day. In most classes, if I'm waiting for an answer to come,

I need to stay silent for a good five or ten seconds before the first brave soul kicks off a content-related discussion. This time was different. I waited minutes. It was an uncomfortable feeling.

I joked, "Y'know, if you ever want to have a seat in this class, you're going to have to figure out what this room looks like and plop some down."

More waiting.

Finally, one of my students huffed, dropped his backpack to the ground, and marched to the back of the class. He hefted a desk and brought it to the front of the room. "C'mon, guys. Let's put these desks out."

Students half-heartedly shuffled from the front of the class to the back, each moving a desk and then a chair. They arranged a confusing cluster of desks in two arching rows facing the whiteboard.

With minutes of class time left, the students sat down glowering at me, seeming to say, "Look, we took your desks from the back of the room and set them up for you. Happy?"

A lot didn't happen on that first day of school. I didn't establish my norms and expectations for the classroom. I didn't explain how design affects what happens in a classroom. I didn't consider how the context of being new students in an already overwhelming environment affects the contexts of learning. I screwed up.

Taking these factors into account, the opening classroom design activity that I created for my students did the opposite of what I hoped it would. My students ended up recreating the traditional view of what classroom spaces are supposed to look like. With rows of desks facing the front white-board, my classroom didn't look different from what they knew classrooms are supposed to look like. Even though this was not the best start to the year, the process still created a space for iteration and a tone of collective experimentation in the class. The process still built community. Discussing the incident later in the week, the students knew I was trying something different and that it didn't go as planned.

From this rough start to the year—trying to do something transformative but primarily reinforcing student assumptions about classrooms and diminishing their confidence in me as their teacher—our yearlong inquiry into a wireless classroom emerged. This first-day-of-the-year fiasco highlights how mistakes can still sustain useful community-building opportunities and how efforts to develop an ethos of making within my classroom began.

This was the first of several steps I took to discover what a contemporary teaching philosophy might look like when it accounts for the cultural changes that have arisen with advances in technology in recent years. Moving desks might sound counterintuitive to the picture I painted above: we didn't yet have any of the flashy digital devices discussed in previous chapters. And yet designing, collaborating, and opening up the design and ownership of classrooms are precisely what investments in technology should be trying to accomplish. Traditional English activities, for example, are simply another extension for encouraging students to guide instructional contexts in the classroom. Civic lessons must engage young people and must not be boring.

Making Sure Civics Isn't Boring

Based on my experiences with the model of change examined in chapter 4 (inform, perform, transform) and the leadership skills that emerged within my classroom, I believe that teachers can be a revolutionary force for social change when our practice adjusts to the contextual environments in which we find ourselves. For example, one of the guiding theorists for how I developed my teaching practice is Brazilian educator Paulo Freire. A teacher first and foremost, Freire's theory of a "pedagogy of the oppressed" emerged from the time he spent as a literacy educator in rural areas of Brazil where he helped adults to read and communicate, to understand root causes of inequity within their community, and to take action to address these oppressive conditions. Politically charged, Freire's theory is intentional about developing the critical consciousness of learners and for them to name (or rename) the world around them. For teachers, this "problem-posing" model of education is built on a theory of *praxis*. Praxis—the process by which a lesson is enacted—hinges on theory, practice, and reflection (this last part too often is disregarded or forgotten).

Although Freire's model is often discussed in theoretical terms, his work is about centering social change and civics within the learning process. There are specific practical ways that problem-posing learning and teaching can be enacted. For example, the conversations within my classrooms centered on cultural and social artifacts and questions related to them, which was a deliberate attempt at creating what Freire called *culture circles* for my classroom to engage in.[15] Freire emphasized how learning can be tailored for transformative purposes in one's local community. But Freire's adapted pedagogy was developed and implemented in a specific context and for a specific time. As much as Freire's critical pedagogy informs how I think about my own classroom, much has changed in society in the past fifty years and within the localized context of South Central Los Angeles. Writing with Donaldo Macedo about literacy, Freire warned against the constant attempts by educators to adapt his approach to literacy within different contexts:

> I could not tell North American educators what to do, even if I wanted to. I do not know the contexts and material conditions in which North American educators must work. It is not that I do not know how to say what they should do. Rather, I do not know what to say precisely because my own viewpoints have been formed by my own contexts.[16]

Arguably, Freire's sentiment here can be applied to policymakers who dictate the kinds of instructional decisions that teachers should be making. Because the contexts of classrooms change from year to year, telling teachers what to do seems to ignore the expertise we each carry into our classrooms.

Ultimately, the key tenets of Freire's work (and the larger scholarship on critical pedagogy)[17] must be constantly adapted, questioned, reimagined, and transformed. Looking at the contexts of learning in my classroom and offering some initial guidelines that could be applied to other classrooms, I've reimagined a vision of civic engagement that responds to student interests, teacher knowledge of what happens within the classroom, and instruction that engages students in the present civic needs of their local communities. I see this work as tied to a critical pedagogy that is transformative for the school community, that engages students and their interests, and that mirrors learning that is connected in today's globalized society. Pointing to both the ubiquity of mobile technologies and the fact that updated teaching practices should not rely on them, I call this model a *wireless critical pedagogy*. In this model, the word *wireless* can represent digitally advanced classrooms (the mobile-media-enhanced classrooms) as well as technology-free classrooms (classrooms that do not rely on technology for transformative learning). Classrooms should respect the key participants in these spaces—students and teachers—and be less influenced by the policies, supplies, and "stuff" of classrooms that too often get in the way of learning. Likewise, adults need to treat students in these spaces in ways that foster civic-based learning opportunities. In looking at some of my study's missteps, it is clear that adults in schools have a lot of learning to do.

"Like Reading"

Although this chapter focuses on civic learning and the roles played by making and by being flexible in the context of a large urban public high school, my focus as a researcher and an educator is often on the literacies that young people develop. In the past three decades, this literacy field has seen tremendous shifts that include expanding literacy research to literacies research, which has emphasized that we interpret, produce, and communicate in modes and genres that can vary greatly. Further, a group of literacy researchers called the New London Group expanded this language in 1996 by illuminating the idea of "multiliteracies."[18] Like civics, literacies shift alongside cultural and social changes over time, so that as our relationships with technologies evolve, so too do the kinds of literacy practices we engage in.

The initial writings of the New London Group were published more than twenty years ago, but today, the notion that students engage in literacies that are varied and extend far beyond "words on paper" is a well-established tenet of educational research. Even so, understanding what new forms of literacies mean in a wireless era and how these literacies shape civic learning still are largely not understood.

Recognizing that reading can mean different things depending on the contexts and outcomes that teachers expect, I provided significant redundancy of the key text for the class. Because the novel we read that year, Cory Doctorow's *Little Brother*, is a freely circulated book with a Creative Commons license, I provided students with several choices for how they could access the text. Some students took home paper copies of the book, and others opted to read the digital version of the book optimized for the screens of their iPods.

Of the many interviews and observations I documented over the course of this study, my conversation with Solomon might be the one that I return to most often. In challenging this chapter's focus on the intersection of civics, literacies, and student meaning-making, Solomon questioned what the act of reading means in the multimodal, plugged-in world that our students are immersed. He forced me to question what counts as reading.

It is nearing the end of May, halfway through our iPod inquiry. As students come into the classroom and the bell rings, I settle into the usual routine of encouraging students to take out their silent reading books and prepare for the usual twenty minutes of sustained silent reading. As usual, earbuds are protruding from some students' ears. I ask Ras, sitting closest to the seat I've settled into, if he would mind turning his music down since others can clearly hear it in the class. After five minutes of the restless rumble of the class, a brief scan suggests to me that all of the students are reading, pretending to read, or in the case of Precious and Solomon, perusing my bookshelves in the back of the room.

I turn my head down and focus my eyes on the academic text I'd brought for silent reading.

Five minutes pass. I look up. I cast a disapproving glance at Ras, who is fixated on a game on his iPod, and he returns his attention to his book. I then look over to see what book Solomon has selected from the class library. Instead of reading, however, Solomon is staring absentmindedly ahead, nodding his head in time with whatever is pulsating out of the single earbud in his ear.

I quietly stand up and make my way over to Solomon, who is inconveniently located diagonally from me. As I approach, Solomon looks at me and nods his head to suggest, "What's up?"

I point to my book and silently raise my eyebrows, as if to ask, "Why aren't you reading?"

Solomon shakes his head subtly and purses his lips, signaling to me, "Naw, ain't gonna happen."

Wondering if I am misinterpreting our silent conversation, I breach protocol and speak quietly, "Where is your book?" This elicits a displeased look from Elizabeth.

Solomon rolls his eyes. He and I both know how this routine goes because I go through the same exchange with Solomon on a near-daily basis. My strategy has not been working, and yet I find myself once again hearing Solomon say out loud, "I told you, mister: I. Don't. Read."

What follows is a vexed conversation with Solomon, who is resolute in his obstinacy. To Solomon, the issue is cut and dry: reading is boring. Period. Just because he has demonstrated in the class that he *can* read does not mean that he *should* read.

Twenty minutes later in class, we are discussing the fictional social network created within our class novel *Little Brother*. At one point, I ask students if they are following along with the text.

Katherine says, "It's more interesting when you read it. Can you just read it to us?"

Other students nod their heads in agreement with Katherine.

I respond, "Hmmm. If only there were a way for you to listen to the book being read to you. Oh, yeah. I put it on your iPods!"

Although most student laugh at my sarcastic tone, Ras shakes his head: "No, but that's boring." The audiobook is not a hit with Ras.

Speaking loudly above the general chatter and giggles from the discussion of the audiobook, Solomon says, "No, you just gotta listen. You can't do anything else. It's like reading."

Although I was aware that Solomon was listening to the audiobook, I was surprised by how he viewed the act of audiobook listening. Solomon switched from stating that he does not read to, minutes later, guiding classmates through the process of engaging with work that is "like reading." His comment suggests significant changes in literacy development and the possibilities of participatory media for improving individual student outcomes. Further, both Solomon and I could have acted differently in the face of his frustration with reading. As text becomes a multimodal enterprise—some young adult literature includes links to videos that continue the text's narrative, and many books include additional online and transmedia content—the process of reading and engaging with an author's work becomes much more than turning pages physically or virtually.

The class took Solomon's point seriously, recognizing the kind of full attention that Solomon invested when listening to an audiobook. Further, Solomon's off-the-cuff remark that listening is now "like reading" signaled a shift in the ways that digital media are changing the in-class environments that English teachers face.

For a student like Ras, however, the audiobook did not adequately change the context of reading engagement in my class. The text was still boring. Not all uses of mobile devices or technology will be excitedly adopted by every student. They are not the panacea that some educators assume.

Solomon's comment that listening to an audiobook is "like reading" leads me back to an initial challenge of defining "reading" in the twenty-first century. Because teachers and policymakers lack an understanding of how digital tools change and remediate literacies, students are left feeling that their daily practices lack value and meaning. More important, when we exclude "like reading" practices that students engage in socially—viewing and commenting on online videos, text messaging with peers, and producing original photos, videos, and Tumblr posts—we further isolate school relevancy in ways that severely limit the civic outcomes of learning for young people. The socioeconomic context that we are preparing young people for needs strikingly different preparation than merely the three R's of traditional schooling.

Despite all of these global changes in how we interact with and socialize through media, English classrooms still use tools that—although still foundational—disregard the kinds of products students will be expected to interpret, produce, and improve in the future. After class, I reflected on how Solomon's insight challenged my own thinking. When I was vexed by Solomon's unwillingness to read, I was vexed at the disconnect between how my classroom reinforced traditional forms of reading in an era when participatory media was not only the lingua franca of students out of school but also a necessary component of twenty-first-century social engagement. For Solomon to be "like reading," he needed to engage with texts in ways that looked beyond traditional classroom practices. Although he stated that you can't do other things while engaged in this activity ("you just gotta listen"), he multitasked, often listening to his audiobook while texting and engaging in various activities on his mobile device. "Like reading" does not require the total engagement that teachers traditionally expect or demand in a classroom.

Part of the "like reading" challenge that Solomon illustrates is that if we use mobile devices only to replicate existing literacy practices, we squander school financial resources and falsely convey that digital tools are tied to old paradigms of engagement. During the same year as this study, SCHS invested heavily in placing a SMART Board (an interactive white board) in nearly every classroom on campus. As a teacher known for using technology within my classrooms, I helped many colleagues plug in their computers and tap on the large board to get acclimatized to the new digital device. And although I helped many teachers figure out how to "use" SMART Boards, I also recognize that most of them would end up using them to replicate what was already being done on whiteboards, which, in turn, replicated what was already being done on chalkboards. If technology is not used to push forward new methods of engagement—to help illustrate new ways to be "like reading"—it is a financial sinkhole. We must question what will be gained from monetary investments when we discuss literacies and access in schools.

Solomon helps us question what "counts" as reading and broadens this conversation to consider the flexibility of schools and their relationship to society. His personal interests are linked to the community's needs in a new and contextually different "wireless" society.

Counting Texts and Socializing the Mobile

For the past decade, I have kept a log of each book I've read,[19] mainly to assist my imperfect memory and to preserve and reflect on what each year's reading yields. The books are a diverse crop in both genre and mode. A healthy portion of the books are print books (the kind that are heaped on my desk and night stand). A few of these books are bound comic books, vibrant in color and plot and quickly read. An increasing number of these books are digital books that I read on tablets while traveling, lightening my luggage by a pound or two. And a robust handful of the books are audiobooks that I listen to while multitasking, as Solomon did.[20] A few friends have protested that audiobooks don't count: "You didn't read those." But again, what counts? For my own list, I include books on my log for retention, not because I physically completed the task of having my eyes track a prescribed number of words on a page.

We tend to consider learning, engagement, and participation in metrics of the past. Our discussions of twenty-first-century skills and tools rely on a vocabulary that understands reading, writing, and learning in schools in ways similar to how they have taken place for over a century. But society is fundamentally different today than it was in the past, and it is always changing. Staying stagnant in our practices means letting cultural practices surpass classroom-based civic lessons for students. Bringing in youth expertise and youth passions within classrooms is crucial. Without collectively working toward a responsible, humane use of mobile media and its culture in schools, our in-school mobile media practice will perpetually be out of sync with the real-world experiences for which we are preparing youth. Wireless forms of engagement in classrooms must push for an empowered identity for young people in relation to their mobile devices. Through continual and responsible use of these devices within classrooms, students can enter universities, job settings, and the public sphere with an orientation toward a purposeful use of mobile media.

Although I may have mishandled asking my students to design the arrangement of desks in our classroom, the activity highlights two overarching beliefs I hold about classrooms, civics, and community:

- These spaces should mirror the types of collaboration and learning found in the "real" world.
- They should allow students to have fun.

This second point may seem somewhat surprising, if you think that school isn't supposed to be a place of fun. But if we cannot support the interests and passions of young people, we are deadening their enthusiasm for learning with each minute they sit in our classrooms. Educational applications of technology, of texts, and of new policies must adhere to both of these components, or we need to reconsider our assumptions about the role of learning in schools.

Conclusion: Hacking Civic Learning

Looking at how students socialized, played, produced, and communicated via the mobile devices in our classroom, it was clear to me that a powerful academic use of technology hinged on one key factor—mutual trust between student and adult. Minerva's struggles to write in the class and Dante's investment in gameplay and technology, discussed in previous chapters, reflect the fact that civic and individual changes come through trust and not through digital devices. As indicators of how larger classroom structures are supporting student agency in nonprescriptive ways, the nuances of learning that Minerva, Dante, and Solomon experienced highlight possibilities. The moment-to-moment ways that experiences can encourage engagement beyond the walls of the school are the civic potential found in classroom trust. When trust is lost, as was the case when substitute teachers taught in my classroom, technology can do little to improve the status quo. Rather, as tools of amplification, these devices can exacerbate conflicts between adults and students. Recognizing that relationships and trust between teachers and students are at the center of powerful models of school learning, we must consider the kinds of learning and political identities that are made within classrooms each day.

Instead of classrooms that replicate traditional models of out-of-date education where teachers fill students with the knowledge they need to pass exams and matriculate into their next set of courses,[21] we can imagine classrooms that function as hackerspaces (also called makerspaces). These guildlike spaces provide members with tools (including computers, three-dimensional printers, hammers, nails, and power saws) and spaces in which to collaborate, build, tinker, and invent. Hackerspaces have spread in popularity across the globe in recent years. They are not purely workspaces but are hubs for fostering community. Hackerspaces are filled with possibility, driven by peer motivation, and prized as places where adult and youth-centered connected learning occurs in modern "affinity spaces."[22] They are spaces where ideas are shared and contested and where civic identities are sculpted.

My own classroom lacked power tools and 3D printers (and even consistently functioning 2D printers), but mobile devices repositioned it as a space for collaborative and individualized engagement and production. My students' iPods, through the careful support of a classroom community, afforded them the opportunity to engage more humanely within an institution that may have treated them otherwise. For example, in framing my own pedagogical beliefs about these wireless tools, I recognized that these devices engendered empowered identities in classrooms. Through connecting devices that carried youth-invested cultural capital within the classroom, iPods made what is often seen as an uninviting environment into a space of familiarity. For educators, this is a key shift: rather than seeing culturally relevant curriculum as simply the media used in classrooms (such as popular music, films, and culturally responsive texts), the *medium* used within classrooms can be a critical component of engagement. At the time of my study, iPods were a youth-validated tool that I tried to harness for my own critical pedagogy. However, merely coopting the interests of young people is not sufficient.[23] Devices that are popular with students are not necessarily appropriate for use within classrooms. What is important is the context of the classroom community and the ways they are included within the classroom.

When the individual production moments and peer-to-peer collaboration seen in this chapter are coupled with the larger acknowledgment of how my classroom could be seen as a broader learning ecology, the networked value of civic learning becomes somewhat clearer. This is not about teaching civics but about learning English language arts content in ways that support civic identities, questions, and engagement.

Looking at the aggregate of minor moments of engagement highlighted by a student like Solomon reminds us that *schools don't have to look like what we think schools should look like*. From rebuilding classroom design, desk by desk, to reframing what counts as reading, civic learning today requires reimagining what schools look like and what happens inside these spaces. Instead of a class being something that a student like Dante passes or fails, it might be a place where students construct narratives and make gameplay decisions based on the choices, expectations, and guidelines provided by their teachers. Students' agency to shape the direction of a class might previously have been stifled, but new pathways can be delineated. Writing previously about the possibilities of gameplay and storytelling, I discussed how storytelling and narrative can essentially remove failure from the vocabulary of educators: "When we shift the context of learning in a way that engages youth in storytelling, there is no longer a need to frame things in terms of failure. Failure ceases to be a concern when children are emboldened as storytellers. Instead of failing, storytelling allows narrative to help guide proficiency."[24]

Further, when the classroom becomes a space that includes more than the usual definitions of pass and fail (even if these two terms are tied to end-of-the-quarter summative assessments), we can reimagine learning as what John Seely Brown has called a "learning ecology."[25] As a space that shifts from "using technology to support the individual to using technology to support relationships between individuals," a learning ecology "is basically an open, complex adaptive system comprising elements that are dynamic and interdependent."[26]

Considering that the goal is for students and teachers to thrive within these ecologies, these spaces must become spaces for peer-driven leadership and engagement. Rather than having a teacher drive real-world contexts of learning, students can shape the sphere of engagement in the classroom in ways that simulate public participation. Further, the work produced in this ecology becomes positively entangled in the world beyond school. These ecologies of making and learning are—as discussed at the beginning of this chapter—civic spaces. My vision of classroom learning is one that makes schools into vibrant spaces that can guide the actions of students like Solomon outside of schools. Teaching is about substantiating a "civic culture" in classrooms:

> a continuum, ranging from organized public activities and associations of various kinds (which might include groupings based around music, sports, or cultural interests), through "parapolitical" activities such as campaigning, volunteering, and community activism, to politics in the more formal, official sense of parties and governments.[27]

As the contexts of civic culture become more messily combined with other forms of engagement, educators must look to how their classrooms support the dispositions for students to transform and lead as part of their routine public, civic life.

Conclusion

Welcome to C:\DAGS

It is a Saturday morning in a school gymnasium—a little earlier than teenagers normally wake up on weekends. Music thumps in the background, and seats are grouped facing a brightly lit scoreboard. Over a dozen teams of students are spending their entire weekend playing, designing, and collaborating at the Critical Design and Gaming School (C:\DAGS), and I am standing in the middle of their first schoolwide game jam.

At the same time that I was analyzing the data collected in my classroom and shared in this book, I was also part of a collective of teachers, parents, students, alumni, and community partners from the South Central High School neighborhood who were designing new spaces for learning within the large Los Angeles Unified School District. As a result of these labors, three connected sister schools now share a new campus just a couple miles away from SCHS. The Schools for Community Action (SCA), housed at the Augustus Hawkins Learning Complex, are organized around principles of renewal, community support, and justice. And although these schools went through an official approval process with LAUSD and are public schools serving their local community, they do so on a foundation of five core values[1] developed by the design team of teachers and community members:

1. **Student centered:** We believe that education should always begin with a strong respect and understanding of each student's potential and desire to learn.

2. **Community collaboration:** We believe that authentic community collaboration leads to transformative school design.
3. **Innovation and excellence:** We believe that teachers should constantly improve their practice to ensure students achieve new levels of success.
4. **Social justice:** We believe that our community deserves better educational opportunities than have been historically provided.
5. **Sustainability:** We believe in creating interlinked strength between the three small schools of the Augustus Hawkins campus.

Built on these core values and theories of connected learning that take into account many youths' interests in technology and gaming, C:\DAGS is a high school that incorporates the research on play, design, and community development that I explored in my classroom. C:\DAGS was built from some of the key ideas I've shared in this book as well as ideas and innovations from several of my teaching colleagues. This is not the only school to offer such curricular and structural choices. Quest to Learn and other schools opened by the Institute of Play have focused on this area recently as well.[2] Along with C:\DAGS, the other two schools at the Augustus Hawkins Learning Complex are engaged in powerful forms of learning in rich and thematic ways. The Community Health Advocates School (CHAS) pushes for understanding community health in authentic learning contexts, and the Responsible Indigenous Social Entrepreneurship (RISE) School offers opportunities for understanding what business and exchange mean in South Central High School.

Seeing many of the lessons I learned through my research being enacted during the weekend-long game jam was thrilling. Everywhere I looked, students were engaged. This was not a quiet space. The C:\DAGS game jam was an astounding display of youth ingenuity. It included a frenetic Connect Four competition that transpired throughout the weekend, a competition (with updated scores for each team) to develop a playable game by the Sunday afternoon deadline (both digital and analog teams were competing), and the creation of a lounge from stacks of boxes where students could escape the design fray. More important, however, was the emphasis in the game jam on community development. Teachers, students, and community members worked together and embodied an ethos of exploration and playfulness that isn't regularly seen at the schoolwide level. Although I had cultivated spaces for students to share vulnerability, to socialize in ways that reflect the "real world," and to transform classroom learning into something that could sometimes be untethered from textbooks and graffiti-scrawled desks, I was seeing this approach adopted in a more widespread manner during the game jam. Even more important, the ongoing teacher-led professional development that organized this game jam meant that these classroom teachers were working at continually improving the practices I'd explored in my classroom.

Just weeks before this first game jam, the three Schools for Community Action celebrated the graduation of their first class of seniors. Although LAUSD and other large districts wrestle with the challenges plaguing schools around academic success, game-based approaches to instruction, learning, and civic support are proliferating vibrantly at this public school in the heart of South Central.

More Changes as Usual

The opening of C:\DAGS wasn't the only significant change I saw in the Los Angeles Unified School District in the few years after I completed my study. LAUSD also infamously rolled out a more than $1 billion initiative to put an iPad in every student's hands and to increase wireless and other infrastructural support for the thousands of students in the city. The district initially paid more than the price that these tablets retailed for in stores because they were preloaded with educational software, which was not necessarily aligned with individual teachers' goals. In this sense, the mobile devices being handed over to students were not about powerful instruction but about powerful control and data collection.[3]

The rush to get these devices into schools in the fall of 2013 illustrates the numerous assumptions that were made at LAUSD and nationally. Reflecting on the roll out, *Los Angeles Times* educational reporter Howard Blume begins a September 2013 article with this comment:

> It took exactly one week for nearly 300 students at Roosevelt High School to hack through security so they could surf the Web on their new school-issued iPads, raising new concerns about a plan to distribute the devices to all students in the district.[4]

Writing for the blog DMLcentral, my colleague Thomas Philip and I expressed our surprise that it took as long as a week for such a hack to take place: "These students' acts of subversion bring the real crises in schooling and learning to light. Educational equity in the city's poorest schools cannot be bought with a hefty check to Apple."[5] Although the deal was temporarily halted amid claims that Apple and testing company Pearson received preferential treatment (allegations that helped lead to the district's superintendent resignation), all districts can benefit from looking at the attempted rollout.[6] As Philip and I note:

> LAUSD's highly publicized missteps reflect problematic trends in education across the nation. When it comes to technology in the classroom, policymakers and administrators chase two irreconcilable objectives: nurturing students' ingenuity and tussling to control young learners. They desire creative, inquisitive youth who access a world of digital information and teachers who skillfully facilitate authentic, individualized learning for all students. Such a model of learning is based on trust in students and trust in teachers—areas in which District officials consistently score "far below basic."

LAUSD wasn't the first place that one-to-one tablet programs were implemented in the United States. But the hefty price tag and size of the district (it is the second-largest school district in the country) made the LAUSD missteps an important lesson for technology evangelists in districts across the country. The attempt at iPad implementation was not seen as a warning sign about our school systems' reliance on digital technology for humanizing education but more as a business-as-usual debacle that was endemic of sprawling district mismanagement.

LAUSD's blunders aren't a sign that all technology is destined to fail when implemented at a large scale, but they are a warning sign that the purposes of implementation will drive innovation and learning. If relationships are at the heart of learning—as I and many sociocultural educators believe they are—then we must continually ask questions about the kinds of relationships that new technologies foster and enable. Most important, the relationships between youth and teachers are a key relational component of schools. Do we insert technology into classrooms in ways that strengthen these relationships? Further, what kinds of relationships do we want students to have with their devices? Unfortunately, these kinds of questions are not usually at the forefront of district administrators' minds when considering the adoption of one-to-one devices in our current era of high-stakes testing and evaluation.

The fallout from LAUSD's botched attempts at districtwide implementation continued for months.[7] And although researchers, media pundits, and politicians debate the program and bandy about words like *efficacy* and *fidelity*, we are placing children into an environment of continual upheaval of order. As educators scramble to figure out how to meet students' needs while incorporating new devices and new software, the change-as-usual approach is affecting the learning and expectations of students on a day-to-day basis.

At the same time that LAUSD's approach to technology integration placed it at the center of national scrutiny about the uses of technology in schools, other significant changes have been occurring nationwide. The Elementary and Secondary Education Act (ESEA) of 1965 has been renewed every five years since being enacted during Lyndon B. Johnson's administration. Recently, the Every Student Succeeds Act of 2015 largely extends some of the work enacted under the No Child Left Behind Act of 2001. (Some positive changes implemented in these acts are beyond the scope of this book.) Likewise, since 2010, the Common Core State Standards have been adopted by forty-two states (as well as the District of Columbia, the U.S. Department of Defense Education Activity, and four U.S. territories), including California (and thus SCHS).[8] The organization of standards in English language arts classrooms into four distinct strands has unified conversations and expectations within classrooms.

Ushered in alongside the U.S. Department of Education's Race to the Top,[9] the Common Core State Standards also meant that states would measure their students' annual academic performance through either the Smarter Balanced Assessment Consortium or the Partnership for Assessment of Readiness for College and Careers (PARCC). The data on student performance on these tests from state to state is still being collected and analyzed. More important, the language on these exams reveals certain expectations for kinds of student learning.[10] These exams require students to utilize technology. A testing literacy that once required students to use a multiple-choice process of elimination now asks them to consider where cursors are tapped and how to enter text into the appropriate on-screen boxes. Assessment, then, becomes even more entangled in technological uncertainties. In 2017, the newest U.S. Secretary of Education, Betsy DeVos, makes federal decisions despite having no previous experience in public schools, highlighting future uncertainty about the direction of educational policy.

Students in schools in the 2010s have experienced constant changes in their classrooms. Standards have led to new kinds of expectations for students' academic learning time. Assumptions about the role played by technology in classrooms have meant that educators may struggle with understanding and implementing district assumptions. The social components of mobile media are treated with skepticism and wariness. Filters are still in place, and student-owned mobile devices are still frequently confiscated. The tools and language of classrooms are changing even while the power structures and expectations continue to remain the same. What kind of academic journey are we sending our students on? Perhaps a turbulent one.

Doing More Now and Doing More Tomorrow

At the end of every school year, I feel that I could have done much more in my classes. The content we explored conformed to the school's course objectives and standards for each class, but I also think of the missed real-life examples, texts, and engaging activities that might have drawn in one more student or made learning even more memorable for students. As an educator, this gentle self-assessment at the end of each year also comes with a renewed sense of commitment to teaching. Improving my practice as a teacher comes from teaching more and reflecting on the pedagogical risks and detours I take.

In her book *Teaching Community: A Pedagogy of Hope*, bell hooks writes:

> Whenever I was frustrated with the stale, unproductive, deadening energy in my classrooms, I could usually shift the mood by threatening to give my "this is our life" lecture. The one that begins with death and dying. It is a small talk about the quality of life in the classroom, a reminder that our time together can be utterly satisfying, complete, a space where we can lose all thought of the future.[11]

Reading hooks's language, I recall my own version of the "this is our life" lecture that I've given in high school and university classrooms. The emphasis on the present moment is tricky when talking with a generation of learners who are infatuated with it. Facebook updates, online check-ins, selfies: these are examples of *existing* in the moment rather than *experiencing* the moment, a challenge that media scholar Douglas Rushkoff describes as "present shock."[12] With youth spending so much energy sharing while taking selfies and "liking" the activities of friends in the present moment, Rushkoff suggests that they are unable to experience fully the time they are immersed in.

Later in her chapter, hooks adds to her focus on being mindful of the present:

> Teaching students to be fully present, enjoying the moment, the Now in the classroom without fearing that this places the future in jeopardy: that is essential mindfulness practice for a true teacher. Without a focus on the "Now" we can do the work of educating in such way that we draw out all that is exquisite in our classroom, not just now and then, or at special moments, but always. Teaching mindfulness about the quality of life in the classroom—that it must be nurturing, life sustaining—brings us into greater community within the classroom. It sharpens our awareness; we are better able to respond to one another and to our subject matter.[13]

As I look back at my class of ninth graders and the lessons we codeveloped around mobile media in classrooms, I am reminded that the key experiences weren't found in the apps students downloaded, in the traditional writing and reading tasks I assigned, or even in the gamelike activities of Ask Anansi. The moment-to-moment, transformative memories for me and—I expect—for my students were the spontaneous times a classroom member experienced enchantment, surprise, *connection*.

Across the many interviews, observations, and teaching that were a part of this year-long study, the most prevalent finding that I return to is that adults and students spend a lot of time misunderstanding one another and establishing power hierarchies to thwart human connections. Learning in schools does not happen when the time for youth socialization is disrupted, youth and adults are distanced through school policies, or mentoring relationships are stifled by classroom squabbling. Learning takes place during the civilizing and mindful opportunities when students and teachers connect and laugh and understand one another. At the same time, the "work" of schools doesn't look much like how people work and interact in the real world beyond the gates of South Central High School. As members of the teaching profession, we can do more for our students by mirroring the social and academic practices that happen in our daily, nonschooled lives. And we can do more by ensuring that such applications are engaging and fun. This is how we imbue *good reception* for youth in the civic spaces beyond schools.

Looking at how students are changing culturally both inside and outside schools, we must look to new ways to instigate delight in inquiring minds. As such, throughout this book, we looked at the macrostructures of learning and technology in a school like SCHS but also followed the specific moments of learning for students within my classroom—Minerva's writing in chapter 3, Dante's competitive spirit in chapter 4, and Solomon's "like reading" relationship with technology in chapter 5. In all of these explorations, I am reminded that although researchers like myself spend a significant amount of time staring at and scrutinizing youths' relationships with blinking and chirping devices in classrooms, these devices should be designed to augment *relationships*. Rather than look at the technical affordances of the latest iDevice, we should consider how technology slots smoothly into aiding teachers and students in sharing their stories, their beliefs, and their voices. Hope, fear, love, *action*: these are the feelings that shape adolescence resilience.

Yes, technological developments and customized gaming experiences can help learning happen in classrooms, but they cannot be the focus for these spaces. Relationships must always reign supreme.

Not Just Telling Teachers What to Do

Adults—security officers on the tardy lines at SCHS, substitute teachers in classrooms, concerned parents, policy makers, and often imperfect educators in classrooms—are responsible for a lot of mistakes in shaping how young people see themselves as learners, civic agents, and members of society. A lot of other books complain about the problems with teachers today. These teacher-blaming books detail how to "fix" the current crises in classrooms (and are often written by pundits with little in-school credibility). As Anthony Bryk and his colleagues note in *Learning to Improve: How America's Schools Can Get Better at Getting Better*, "It is hard to reconcile, in a nation that spends as much as we do on public education, that some students predictably will fail, regardless of the resources and attention directed to them."[14] The typical response, these authors explain, is to blame teachers, a dangerous example of attribution error: "When we see unsatisfactory results, we tend to blame the individuals most immediately connected to those results, not recognizing the full causes."[15]

As a teacher, I spent a lot of time sitting in mandatory staff meetings being "professionally developed" by well-paid professors, researchers, and consultants. As a nation, we spend enormous amounts of money telling teachers what to do. Rather than dictating to teachers what to change, I have invested my time as a researcher and teacher educator over the years since I left my SCHS classroom documenting what models of connected learning look like in classrooms today.[16]

Three key interrelated concepts often are missing from reform discussions about the needs of the teaching profession:

- Technology will never fix schools.
- As technology shifts, culture shifts alongside it.
- Schools must adjust to the cultural shifts that affect how young people learn, not to the technologies that are released from year to year.

In looking at these shifts, the role played by teachers must again be emphasized. When designing a curriculum that revolved around mobile-media use in my classroom for this study, I focused on features and applications that were intuitive and already in use by a mobile generation.[17] In this way, my classroom reflected the cultural shifts within U.S. society. By selecting the basic functions that are featured on most mobile devices, I wanted to ensure a universality of applicable use in most classrooms across America. I focused on the most basic features of mobile devices with two goals in mind—to ensure student familiarity, comfort, and confidence with how these devices were used within my own classroom and to alleviate teacher discomfort with utilizing these devices in settings beyond the scope of this study. Based on my experiences with my peers at SCHS and in several teacher-education programs, I've seen that the introduction of new technologies and sets of tools into classrooms is often received with trepidation. I wondered if using very basic attributes of mobile media devices might make them more readily adaptable by other teachers after this study was completed.

A recurring theme in my own research is a concern that the label "new media" does a disservice to teachers by dissuading many educators from adopting tools that can lead to powerful learning opportunities. In a blog post, I reflected that "without redefining the terms we are using to describe these tools and student work, digital technologies can actually be perceived as a cult-like sub-genre of the stuff teachers use; it can be looked on with bemusement by a critical mass of teachers as a pedagogical circus sideshow."[18] Teacher use of new media is largely relegated to an ancillary, optional activity. A lack of confidence in understanding and using a seemingly complicated feature of a mobile device is an easy reason for a teacher to shun a pedagogy that includes mobile-media practices. But if teachers have to understand only how to text, photograph, and scan images with a phone (practices that many teachers engage in regularly), the transfer of this study's curriculum becomes a more likely enterprise. Just as the definition of "new media" changes as various media-production practices become enfolded into mainstream use, the charge for educators also will evolve. Rather than teaching teachers how to use specific tools, we must collectively work to acknowledge the contemporary practices that we rely on digital devices for and seek out ways to build these into meaningful classroom interactions.

Where's the Technology?

Long before the advent of mobile devices (like phones, tablets, and netbooks), before "smart" boards[19] (which allow learners to manipulate projected images with their hands and with special styluses), even before the Apple IIE (and its perpetually running copy of the computer game *The Oregon Trail*), researchers were looking to technology to make schools better. In a 1961 *Harvard Educational Review* article titled "Why We Need Teaching Machines," behaviorist B. F. Skinner describes the learning benefits of his mechanical device, which taught skills by regularly reinforcing student activities, thereby covering the same material covered by teachers in allegedly half the time.[20] This emphasis on speed, on "teacher-proofing" student learning experiences, and on a constant appraisal of student performance continues to drive the educational technology sector.

With each advance in technology, the promise of the prodigal son of educational transformation is heralded once again. Projectors, Smart Boards, laptops, and tablets are all here to fix what hasn't been fixed before.

As my students and I explored intentional iPod use in my classroom, I made missteps—as did other adults—that affected student learning. The moments that did facilitate stronger learning among students were moments in which technology functioned in parallel with social and communicative practices already in classrooms. The *context* of learning in my classroom with technology was set long before the mobile devices ever entered the picture.

Building off the trajectory of critical research on the uses of technology in schools by researchers like Larry Cuban and Neil Selwyn,[21] my colleague Thomas Philip and I have argued that technology in classrooms can help support social contexts of learning but that the assumptions by educational leaders that technology can "fix" schools are shortsighted.[22] Pushing for teachers to explore and understand the constraints of flashy new technologies may not sound as rewarding as seeing students swiping and clicking to their hearts' content, but it is the pragmatic step that continues to be missing. To support teachers, we have framed questions for educators to consider asking about the "3Ts" that technology can support—text, tools, and talk:

- **Text:** Are new texts introduced through the technology? How are traditional forms of texts altered by new technologies, and why are they important for students to learn?
- **Tools:** What learning contexts make this tool imperative to students' lives? How does the tool offer ways of collecting, representing, visualizing, analyzing, and communicating information that contribute to improved learning?
- **Talk:** Are the ways that students communicate made more robust as a result of this technology? Are forms of communication broadened in any significant ways? Limited?

These questions help illustrate the kinds of environmental and social changes that technology can instantiate within classrooms, placing the emphasis on the needs and interests of learners. In doing so, they emphasize that schools should move away from questions that essentially ask "Where's the technology?" and more toward questions like "Where's the learning?" If administrators answer the second question simply by pointing to the digital device in someone's hands and not to the learning ecology of students and teachers, then we are not building humanizing and authentic relationships through technology uses in schools.

The Ship of Theseus

After completing his long travels, escaping a labyrinth, and slaying the minotaur, Theseus sailed home a hero. His ship was a beacon of hope, courage, and valor that was kept in the port of Athens for years. As the myth of Theseus grew, his ship remained a testament of ancient Greek strength. And when the ship's wood slowly rotted, decayed planks were cast aside and replaced so that the ship continued to remain seaworthy in the port.

As the years passed and more and more planks rotted and were discarded and replaced, the pile of decayed wood on the shore became an impressive monument in itself.

When the Greek historian Plutarch looked at this ship and its famous pile of decayed wood, he posed the paradox known as the Ship of Theseus: is this ship that floats in the port and that now is made entirely of new wood the "same" ship as the one that Theseus sailed? More important, if another, identical ship were to be assembled from the discarded wood, so that two vessels mirrored one another in the Athenian harbor, are they the same ship?

This rebirth and examination of the old is not constrained solely to the legend of Theseus. A book that looks at museums and preservation, *The Same Axe Twice: Restoration and Renewal in a Throwaway Age,*[23] looks at what counts as genuine: what pith of authenticity is captured as parts of the old are replaced and polished away as fragments of the detritus of time? These are serious questions for the teaching profession.

What about schools? As tablets, phones, and other mobile devices become entangled in the U.S. education system's ongoing infatuation with high-stakes testing, at what point do these devices get us to a place where "the end of all our exploring will be to arrive where we started and know the place for the first time," to quote T. S. Eliot?

A major lesson that I've learned from our nationwide implementation of the Common Core State Standards and the high-profile blunders of LAUSD's districtwide purchase of iPads is that not that much is changing, even when the devices we integrate into classrooms become flashier and more expensive. Reflecting on this, I return again to educational researcher Larry Cuban. Describing the unchanging landscape of schools today, Cuban notes that a "constancy amidst change"[24] has been key to how classrooms have dealt with technology. Devices—from our cloud-connected tablets down to slide projectors—have been ironed into *the way things are done*. Swiping and tapping on a text when reading it, writing an essay response in a digital box, and viewing textbook publishers' online content are, at the end of the day, *still* reproducing the same kinds of learning and power relationships that have existed in classrooms for over a century.

Looking at the birthdates of the students in my current teacher education courses, I was struck by an obvious fact I had missed. At this point, most of the students I work with who are becoming teachers have been schooled exclusively within the policy mandates of the No Child Left Behind Act. For an entire generation of teachers, the system of accountability and high-stakes testing has been their baseline paradigm of what schools look like. *Of course* technology is going to be a part of the expectations of people living in this current landscape, one new piece of lumber replacing a weathered archaic piece but largely preserving the general form and shape of traditional education.

But what if this newly reassembled ship was undocked from our past? We have been anchored by the culture on which our classrooms are presently based. The college students I teach today want to become teachers both in spite of and because of what they experienced in No Child Left Behind classrooms. However, structural changes in how we interact with one another and with media today in our participatory culture illustrate that we are preparing youth for entirely new contexts of work and learning. Simply making software run assessment tests on a tablet without killing trees isn't enough. (And the ecological footprint of producing and powering millions of Internet-connected devices today is itself an often undiscussed environmental cost.)[25] The ships of our present may be made of the same lumber as those of our past, but we have a different path to sail, new labyrinths to explore, and a new generation of teachers who are willing to enfold their cultural experiences into new possibilities for classrooms.

Not all classrooms need to or should play Ask Anansi or create scavenger hunts that annoy on-campus security guards. The context of who students are and what their needs and interests may be shifts from one community to the next. Mobile devices that encourage personalized communication should facilitate a responsiveness to these needs, an ever-changing raft afloat a sea of equity-driven possibility. Together—youth and adults united—we can steward powerful new journeys for how we understand, learn from, and lead with digital technologies and humanizing relationships.

Appendix A: Methodological Statement

This book began as the primary research for my doctoral dissertation, which was part of a larger interest in supporting the needs of students and teachers in urban schools, particularly in South Central High School. As such, it belongs to a tradition of ethnographic research studies conducted in schools and a smaller subset of such studies conducted by classroom teachers. In this statement, I briefly describe how I entered classroom teaching, what led me to begin my research, and what methodologies I employed in conducting this research. In this background description of my positionality, I try to present clearly any biases I carried with me throughout this study as I attempted to offer an emic (insider) look at student life at SCHS in an effort to ensure that I did not "speak for others, but about them."[1]

As the son of activists and educators, I was raised with an appreciation for literature and social justice. As a multiracial, cisgender male, I regularly heard comments from my students that revealed that they perceived me as someone who came from a background that was more privileged than their own. Although my last name signifies that I am Latino, I am a monolingual educator and was often assumed to be white by my students. After receiving a B.A. in English with a creative writing concentration, I began a social justice–oriented teacher credential and master's program, and through this program I connected a passion for education with a foundation in critical theory and critical pedagogy.

Through this master's program, I began my teaching career at South Central High School and was an educator there for eight years, the second half while concurrently enrolled in a Ph.D. program focused on urban schooling. At SCHS—as a member of the School Site Council, a lead teacher for a social justice small learning community, a member of the school's instructional cabinet, and a member of the local teacher's union and in several other leadership roles—I advocated for the needs of the students on the inequitable B track schedule (described in chapter 1). My experiences in these contexts helped me develop a nuanced perspective of how the school's organization affected classroom practice.

In 2007, through a grant from the MacArthur Foundation's Digital Media and Learning Competition, I codeveloped an alternate reality game for helping students increase their environmental literacy skills. The Black Cloud game (described in chapter 4) functioned as a larger portal for understanding nontraditional approaches to critical education and civic engagement. My collaboration on this game became a turning point in my understanding of school policy and educational theory.

In addition to my grounding as an educator and researcher, my personal cultural and racial background shaped the work I did in this study. Within and around my classroom, I was seen as an advocate for student needs. Over years of teaching, various students confided personal information to me, asked for personal assistance, or admitted to feeling intellectually engaged for the first time in my classroom. I write of these discussions not to boast of effectiveness as an educator but to outline ways I worked closely with a student population that came to trust me.

Because of my unique position as an insider in the school's adult population and as someone already familiar with some conceptions of the school's social spaces for students, I was able to gain access for a study that focused on the day-to-day activities within the school. At the same time, gaining the trust of students who were aware of my position of adult authority was an initial struggle. Students sometimes hesitated to answer questions and were initially unclear about whether I was looking at their mobile use as a researcher or as an on-campus adult witnessing school policy violations. However, former students helped me gain entree into the world of youth mobile-media practices on the campus. Seniors I had taught the previous year advocated for me and introduced me to their peers to help reassure students that I was not looking at their mobile practices with a disciplinary eye. Perhaps more important than the issue of gaining access was the need to challenge my preconceived beliefs about student cultural use of media technology on campus. Although teachers, administrators, and students had in the past seen me as an advocate for youth use of mobile media, I also used my role as a researcher to explore critically how mobile devices hinder classroom learning. For this purpose, using disconfirming evidence to analyze ethnographic fieldnotes was key throughout the study. As such, the perceptions of others that I was a technology advocate as well as my own approach to critical theory contributed to "the epistemological and political baggage" I lugged into this study.[2]

One of the biggest struggles I faced in this study was conducting research within my own classroom and within my own school. Although there is theoretical research that helps rationalize the necessary contributions of teacher researchers,[3] I struggled with the day-to-day challenges. In contending that I did not find an optimal balance between my roles as researcher and teacher at SCHS, I want to note that the findings in this book are often informed by my teacher perspective. This conception of power as a factor of both teacher and researcher roles is one that researchers such as Kathryn M. Anderson-Levitt insist must be made clear when conducting qualitative research.[4] As the needs of students within the classroom were not the same as my needs as a researcher, I often grappled with balancing my conflicting roles in the classroom. For example, although I sometimes wanted to stop activities in my class to ask my students for more details or to write down a detailed note about what I was observing, my responsibilities as a teacher meant that I needed to move at a teaching pace appropriate for the students I taught. To this extent, I audio recorded my various interviews and focus groups and also created ethnographic "jottings"[5] of my in-class observations that were developed into longer field notes and research memos at a later point.[6] My insider perspective as someone interacting with the students balanced with my outsider perspective as an adult watching youth cultural practices helped yield the "dualistic approach" that exemplifies ethnographic research. I also shared early drafts of these chapters with teaching colleagues and some findings with students to get their feedback and responses. Revisions to my analysis (such as my discussion of Solomon when he was "like reading") are informed by these conversations. Although my name appears on the front of this book, the research was done with the guidance and intellectual consideration of as many of my students and colleagues as I could involve.

For the key findings around schoolwide mobile-device use, I interviewed students from all three tracks of the school's year-round schedule and in all four high school grades. I started with a predetermined protocol for the interviews and asked follow-up questions as necessary. I also treated my own classes as interview participants, and their voices echoed and usually corroborated the key claims I've made. In addition to classes with my own students, I conducted a total of sixteen focus groups with more than a hundred students. Each focus group averaged forty-five minutes.

For the instructional portion of this study, I relied on backing up my notes with audio recording and used a LiveScribe recording pen to take notes and to facilitate the tracking of key moments of exchange within the class. In order to best analyze and organize these fieldnotes, I initially coded these data using an inductive approach.[7] I looked for repeating words, ideas, and actions that arose from the fieldnotes and coded these patterns. In analyzing the data, I was particularly interested in ensuring that I did not selectively choose only the data that confirmed the value of mobile media. Instead, I continuously sought disconfirming data for the patterns that I identified. Relying on different types of data, I used "converging lines of evidence" to triangulate my main findings.[8] For example, in one classroom, I observed a student discreetly walking into the hallway as though he was looking for a stronger signal for his phone. Later, in discussing phone reception at the school, students shared locations in the school that are known for having strong and weak phone reception, and one student pointed to various locations on campus and shared how to continue to receive messages from peers in these spaces. By looking at the nexus of these various data, I tried to represent the practices and views of students accurately rather than select unique instances within my data that supported a particular point of view.

My main unit of analysis for this coding was the specific speech event that I documented. As described by Del Hymes, "The term *speech event* will be restricted to activities, or aspects of activities, that are directly governed by rules or norms for the use of speech."[9] Relying on the recordings made in class and messages written, sent, and read, I was able to listen to and transcribe student speech and analyze the digital speech that occurred in text messages and e-mails. Speech events are "a key way of understanding written language ... to examine specific observable events and the role of texts in these events. The literacy event becomes a basic unit of analysis."[10] Because literacy events are "dynamic activities" that can include multiple perspectives, purposes, and participants, an understanding of dialogue and text in this study stems from a critical perspective.[11] As a result, the discussions noted throughout this study about mobile media use and gameplay at South Central were analyzed with a discourse analysis that focused on how language as a cultural tool mediates relationships of power and privilege in social interactions, institutions, and bodies of knowledge.[12] For example, students created narratives that challenged the existing discourse of school structures and locations. Dede labeled the sole green shrub on the school grounds as "Captain Green" in a playful way that (she later told me) was meant to make people look at the shrub "in a different way" (chapter 4). This mode of analysis is "an approach to answering questions about the relationships between language and society."[13] As a study that looked at literacies and civic participation from this critical perspective, the data analysis here explored the verbal and online dialogues that were produced in and shaped meaning in my South Central classroom. Because a discourse is "productive" and iterative,[14] dialogue in my class was something that I viewed not as isolated incidents of speech but more as a string of incidents related to student literacy development.

Finally, concurrent to my analysis of data, I was also designing the curriculum and alternate reality game Ask Anansi to implement with students in my ninth-grade classroom. My design philosophy behind this process reflected my beliefs about the need for students to be codesigners of their own civic learning experiences. This curricular process was designed based on my experiences with doing youth participatory action research (YPAR).[15] Working alongside young people in playing, researching, and designing the work was a key consideration in developing Ask Anansi and the in-class activities the class participated in during the study. As a model that accounts for student voice in the spaces of traditional research, YPAR can potentially upend existing research paradigms in disruptive ways that evoke the disruptive shifts ushered in by mobile media and participatory culture.

Appendix B: So You Want to Design a Game

When I share my experiences with the alternate reality game Ask Anansi with educators, some are immediately ready to dive in and run the game with their students and ask how they can get a copy. Others are exhausted even thinking about doing this extra work: "I'm a teacher, not a game designer. It's nice that you were able to funnel your own geekery into the classroom, but this isn't sustainable work for all of us."

With both of these sentiments—the interest in replicating the game and the wary sense that it is not a feasible task—my response is generally the same. Ask Anansi was a game for a particular time, place, and gathering of individuals. I knew my students and the ways this game would fit into our specific classroom community. As I have stated through this book, your own interests, those of your students, and the foundations of trust and community in your classroom will drive what is needed. More broadly, however, a few guidelines might help you get started with game play and game design in classrooms, so I offer a few suggestions and resources below. At the end of this appendix I also offer my original 900-word game design concept for Ask Anansi. You'll notice key differences between the game concept and its actual execution, and I encourage you to read this document as an overview for a curricular unit plan or inquiry challenge. Although this does not spell out each activity or daily lesson of the game, it offers a way to consider the broader steps of designing meaningful play in classrooms. I encourage you to consider the suggestions in earlier chapters to help guide your own thinking about using this document for your classroom.

Additional Resources

Before diving into guidelines for game design, here are other resources I encourage teachers to check out:[1]

- *The Art of Game Design: A Book of Lenses* by Jesse Schell: This is one of the most accessible books on the key processes of game design. The supplemental deck of cards is also a great resource for classroom use.
- *A Theory of Fun for Game Design* by Ralph Koster: This book's title speaks to what it offers—several activities to spark your passion as a game designer.
- *Grow-A-Game* by Tiltfactor: This deck of cards and free app are helpful for instant iteration and idea generation. I regularly use these cards when presenting game design ideas at workshops.
- *The Educator's Game Design Toolkit: A Game Master's Guide and Player's Guide for Teacher Professional Development* by Antero Garcia and Chad Sansing: My colleague Chad Sansing and I developed this guide as a resource for classroom teachers looking to collaborate and play in classrooms. It is structured around several activities that can be done individually or with other teachers.

Start Small

Long before launching Ask Anansi, I instigated playful moments in my classroom (such as the unsuccessful exercise in student-led desk arrangement noted in chapter 5), designed other alternate reality games, and co-ran an after-school gaming club on campus. As you begin to think about gaming in your classroom, it is okay to start small. It is okay to start *really* small. Many English classrooms use trivia games (digital and nondigital) to review class material. Can you consider how such work could become even more playfully tailored for your students' needs? Similarly, could you see ways to use the collaborative story writing game "exquisite corpse" in your class? (In exquisite corpse, popular during the surrealist movement, each participant adds a line or an image to a story without knowing what was written or drawn before his or her turn.)

Your confidence and capabilities as a game designer will grow with time, so don't worry about experimenting and trying small things first!

Think Like a Game Master

A game master (GM) is the person in a tabletop roleplaying game who facilitates play, guides storytelling, and plays and controls any character that is not played by one of the other players at the table. In the classic role-playing game (RPG) Dungeons & Dragons, for example, several players may converse with royalty, seek gossip-like clues in a tavern, fight a pack of wolves, and come face to face with a vile dragon. The characters and the descriptions of the story that moves the players to these incidents are all controlled by the GM (in Dungeons & Dragons, the GM is called the dungeon master). In a recent book chapter, I discuss how teachers can adapt the role of the GM for their classrooms.[2] What's particularly important—as you look at the experiences of Ask Anansi and your own game design decisions—is that you will need to be flexible and respond to the interests and needs of your classroom. But thinking like a game designer and game master are not that different from thinking like a good teacher.

In discussing the needs of GMs in role-playing games, writer and designer Robin D. Laws describes six different kinds of players that are commonly seen around the table—the power gamer, the butt kicker, the tactician, the specialist, the method actor, and the casual gamer.[3] These gaming tropes are not unique to role-playing games and are not limited to the types Laws describes. You can see some students taking on tactician-like roles while hiding clues in Ask Anansi and other students wanting to find all of the solutions as classroom "power gamers." Perhaps the most important type of player for teachers to consider is the "casual gamer." Translating this role for the classroom means considering the students who may not be intellectually roused by alternate reality, technology, or games. There is no panacea, and what enthralls some will not magically transform your entire class.

The secret to good classroom gameplay is flexible teacher game mastering. If something isn't working, you should be ready to adjust and readapt at a moment's notice. This should feel similar to the flexibility to classroom teaching. They are connected practices.

General Guidelines for Design

- Consider your instructional goals. What do you want to convey, and what do you want students to have learned and accomplished by the time they finish your game?
- For longer games like alternate reality games, build a narrative hook. What is the story you want students to experience?

• Keep it simple. You're not George R. R. Martin. You don't need byzantine plot structures and complex backstories. If it's not a simple story, it's probably not a compelling story for your students to unpack.
• Consider the kinds of play students will engage in. Center the story and game around one or two fundamental activities. In Ask Anansi, this was a scavenger hunt. In a previous alternate reality game, it was Capture the Flag. Consider what it looks like to readapt a game like checkers, Crazy 8s, or Hungry Hungry Hippos. (Such adaptations do not have to be literal. Start with a game mechanic, and build from there.)
• Pace yourself. Things usually take twice as long as you intend, so don't be afraid to slow down your design and implementation.
• Embrace the fact that things could (and likely will) go badly. It's okay if things feel like they are falling apart. On the one hand, it's probably not as bad as you think. On the other hand, this is an instructional experience for your next design!

Ask Anansi: An Inquiry-Based Alternate Reality Game of Tricks, Riddles, and Spiders

Overview

Ask Anansi is a community-centered alternate reality game. In this game, students engage in inquiry-based problem solving by communicating with and helping to unravel the stories they are told by Anansi, the trickster spider god of West African folklore.

Anansi, the story-wielding spider god, has answers and solutions to any question students can imagine, and fortunately, these students have recently received a means of communicating with him. Students pose community-centered inquiries to Anansi via his progeny: it's common knowledge that spiders can relate questions to Anansi if you ask them nicely. A classroom spider will help students initially communicate with Anansi.

Anansi's responses, however, are not always clear: he likes tricks, riddles, and befuddlement. As a result, students will require critical literacy skills to unravel the web of Anansi's hints and instructions. Some clues are found outside the walls of the classroom and may appear as posters, barcodes, or phone calls. After a question is asked, it cannot be unasked, and Anansi is known to grow impatient with small children who do nothing but waste his time by not solving his puzzles. Who knows what would happen to their teacher or their classroom materials if they dawdle?

Each Anansi question will take a group effort to answer. However, be careful. Anansi is never satisfied with simply finding the answers to the many questions students ask. He often requires that students work toward resolving the problems they discover.

Concept Outline and Scenario of Sample Play

The goal of Ask Anansi is to guide students toward collective inquiry around a negotiated topic and civic engagement by addressing the underlying causes of these topics. For example, a class may investigate why the food at their school is unpopular. Through doing research on nutrition, budgeting, and distribution of food and by doing qualitative surveys and ethnographic analyses of student perceptions of school food, students may determine that a lack of variety of food due to budget and contracting constraints as well as a social perception that the food is "bad" are preventing students from receiving adequate nutrition during the day. Next, students may determine that one course of action is to develop a coalition of concerned parents and students, speak at school board meetings, and even stage a cafeteria sit-in. Students will reflect on their efforts, discuss changes they have made, and record these steps in text messages, videos, and mapping applications on mobile devices.

Although the main product of this game is problem-positing critical thinking and civic participation, the goal of the game is based in the alternate reality game's fiction: they must satisfy the insatiable need of Anansi for a good story. Asking Anansi a question seems innocuous. A comment is fed to the class's spider to communicate to Anansi: "Why is the food at our school so bad?" A day later, the class receives a cryptic response—the coordinates to the loading dock for the school's food shipments and a riddle guiding them to photograph and blog about the nutritional information they can ascertain.

The game's initial premise of asking a simple question has significant repercussions. Anansi will not simply provide an answer. He tricks, confounds, and teases students. Anansi's messages are often shrouded as riddles, QR codes, or even latitude and longitude coordinates that need to be determined and visited. Like the media messages that students are challenged to assess critically, Anansi's dialogue with students challenges power, dominance, and agency in a capitalist environment.

As students gain more information, Anansi's responses become more demanding. Students will regularly dialogue and blog about their experiences. Anansi may hack or edit their information in an effort to further a good story. After a particularly intriguing development (an interesting plot development for Anansi and a useful analysis of data for student inquiry), students will be given goggles to view a situation through the eyes of Anansi. This eight-eyed mask will guide students to analyze media, their actions, and the information they gain through various theoretical lenses—reader-response, historical, marxist, feminist, postcolonial, psychoanalytic, ecological, postmodern. Students may continue to use these "goggles" in the class long after their initial encounter with Anansi.

After students have completed their initial research and analysis, Anansi tells them that they have the pieces of a great story but need to weave them into action. Students need to work toward a course of action around the information they have received. Collective action and models of engagement are examined by the class, and a strategic plan is developed and enacted.

Students will develop an evaluation mechanism to determine the effectiveness of their plan and will revise with further iterations as encouraged by Anansi.

Anansi confesses at the end of the game to having tricked the students in places with his difficult clues. He suggests that students should recruit others to continue the story they have woven together. After all, Anansi will remind players, a story never really ends. We may continue to tell what happens until the next series of adventures. An inquiry and efforts to improve food quality at a school are ongoing endeavors that students can continue to work on.

Appendix C: A Conversation with Anansi: A Spider's Thoughts on Participatory Professional Development and Alternate Reality Gaming as Youth Participatory Action Research[1]

In early 2012, Antero Garcia sat down and recorded a conversation with Anansi the spider. Over cups of tea, biscuits, and horseflies, Antero and Anansi discussed the alternate reality game called Ask Anansi, participatory professional development, the role of storytelling and gameplay within pedagogical development and teacher community building, and ways to sustain this work in public schools.

Antero: Who is Anansi?

Anansi: Me? Well, that's a long story (and I do love stories, as you shall see). Although many tales have been spun about me, for now it may be useful to know that I am a West African folklore hero and that I often take the shape of a spider (as I do now). I encourage you to read of all of my trickster tales, but perhaps most relevant to our discussion today is the fact that I own all of the stories you can possibly imagine. Getting me to share them with you, however, is another story.

Antero: I heard that one way to get you to share your stories is through an in-school engagement model. What is Ask Anansi?

Anansi: Good question. Ask Anansi is a community-centered alternate reality game. In this game, students engage in inquiry-based problem solving by communicating with and helping to unravel the stories that I—the trickster spider god of Caribbean folklore—tell them.

As the story-wielding spider god, I have answers and solutions to any question students can imagine, and fortunately, these students have recently received a means of communicating with me. Through simple text messages, e-mails, voicemails, and even disruptions within classroom experiences, students engage in a sustained dialogue with me.

My responses, however, are not always clear. I like tricks, riddles, and befuddlement. As a result, students require critical literacy skills to unravel the web of my hints and instructions. Some clues are found outside the walls of the classroom and may appear as posters, barcodes, or phone calls. After a question is asked, it cannot be unasked, and I am known to grow impatient with small children who waste my time by not solving my puzzles. Who knows what might happen to their teacher or their classroom materials if they dawdle?

Each Anansi question will take a group effort to "answer." However, be careful. I am never satisfied with simply finding the answers to the many questions students ask. I often require that students work toward solving the problems that they uncover. And although Ask Anansi operates within a fictitious narrative and the students (correctly) assumed that their teacher embodies the Anansi persona when communicating with them via text messages and e-mails, the gaming environment allows students to act, question, and engage in simultaneously critical and playful inquiry. The main product of this game is problem-posing critical thinking and civic participation, and the goal of the game is based in its fiction: students must satisfy the insatiable need of Anansi for a good story.

Antero: Wow, that was a lot to take in. I'm wondering what the role of this ARG stuff is in teacher professional development.

Anansi: Although alternate reality games (ARGs) tend to layer a fiction on top of a real-world setting, Ask Anansi moves toward participatory learning within classrooms while also guiding teachers toward a participatory pedagogical approach.

This model seeks to inform teacher professional development via direct interaction with students and student expertise. This participatory model draws on Katie Salen and Eric Zimmerman's definition of *transformative social play*: "Transformative social play forces us to reevaluate a formal understanding of rules as fixed, unambiguous, and omnipotently authoritative. In any kind of transformative play, game structures come into question and are re-shaped by player action. In transformative social play, the mechanisms and effects of these transformations occur on a social level."[2]

The shift in focus that happens via transformative social play occurs for both student and teacher. Through teacher collaboration, discussion, and group provocation, teachers move from rote lectures to participatory development.

Instead of simply learning the rules of Ask Anansi and attempting to input them into their everyday practice, teachers come to their professional development space with a set of simple topics or guiding questions they would like to use as foundations for inquiry within their classes. Topics as broad as "How does the Pythagorean theorem affect my daily life?," "How does conflict affect human decisions?," and "What are ways that symbolism affects how I read Shakespeare?" are all suitable entry points for developing a transformative learning experience. The experience becomes less a space for consuming teaching content and much more a generative space: teachers build these questions into a series of areas of inquiry that are fleshed out through student expertise (more on this below).

Perhaps one thing that teachers should keep in mind while developing their lessons is the role that Anansi will play. How will students communicate with Anansi? How will Anansi, as an outside agent, help provoke, move content forward, and drive students toward understanding and content mastery? Although the products that students create and analyze speak to the transformative power of gaming, these activities function within this larger pedagogy of transformative social play. The social space of the classroom and of student experience beyond the walls of the school drive the learning environment. But then again, what do I know? I'm just a spider.

Antero: So that sounds really good, but what are the professional development design goals?

Anansi: Sorry, I don't really know what you mean.

Antero: I guess I mean, this sounds like an enriching classroom activity, but how does it affect professional development?

Anansi: Why didn't you just say that? For teachers, this is an opportunity to do a couple of things. First, it allows teachers to expand their practice beyond the walls of their classroom and to encourage student expertise to guide the work that occurs. To get to this kind of activity, teachers have to shift from roles as experts to roles as co-constructors of knowledge. Teachers need to create spaces for young people to have the space to ask questions in. To do that, teachers first need to be in a space to ask themselves questions within a peer network. The same way you and I are in dialogue with sustained focus on a given topic, teachers will need to explore their pedagogical goals and look at this as a journey that needs to be constructed pedagogically.

Antero: And how exactly do you get teachers to start allowing students to ask questions?

Anansi: Funny you should ask, since you seem to be doing a fine job asking questions here. From my experience as a storyteller and a community rabble-rouser, I've found that people start engaging when they have specific roles to play and spaces within which to ask questions. This question-and-answer conceit, for example, is bounded by superficial constraints that limit us to discussion about participatory professional development. If your role as the questioner were unbounded, we would be talking about favorite pizza toppings and Russian literature. However, by mutually agreeing that we will focus on the topic of participatory professional development and my role in an alternate reality game, we move toward ever more specific learning contexts.

By engaging teachers in an alternate reality gaming model of instruction, Ask Anansi seeks to move student and teacher interactions toward a model of mutual investigation. It is an iterative youth participatory action research (YPAR) engine. By asking students and teachers to collaborate in developing a research question and using the fiction of communicating with me (a story-telling spider), students are encouraged to explore and review their community while teachers engage students as co-researchers through a process of media production and play. This is also mirrored in the participatory professional development. Although teachers may not feel it necessary to communicate with me, this game essentially models the student experience: it is generative through question-driven inquiry.

Antero: Sorry, "youth participatory action research"?

Anansi: I glossed over that one, didn't I? Youth participatory action research is an emerging research methodology. It builds on the principals of participatory action research (PAR) by involving young people in the process of knowledge development. For teachers, it can be a shift into thinking about co-constructing learning experiences. And this professional development model, in its dialogic form, compels educators to move from telling to asking.

In any participatory action research projects are guided by three dominant principles: "(1) the collective investigation of a problem, (2) the reliance on indigenous knowledge to better understand that problem, and (3) the desire to take individual and/or collective action to deal with the stated problem."[3]

This method of instruction—involving students in designing and exploring meaningful learning experiences—"contributes to a way of thinking about people as researchers, as agents of change, as constructors of knowledge, actively involved in the dialectical process of action and reflection aimed at individual and collective change."[4]

Antero: This is a lot of questions! Does this game ever end?

Anansi: I'm really glad you asked that. You see, Ask Anansi helps educators adapt and revise their game and storytelling experience over time. By iterating through various versions of this game (either with the same group of students or a new class), the teacher responds to student interests, specific content goals, and the changing world in which the game and school exist. In terms of assessment, Ask Anansi evaluates student learning through a model I call inform, perform, transform. The three phases of this performative evaluation model expand assessment to adhere to a liberatory model of inquiry-based youth research as follows.

In the inform stage, students gather, analyze, and collate information in order to produce their own, original work. In Ask Anansi, students furthered three types of knowledge during this phase—indigenous expertise of their communities; conceptual understanding of the function of gameplay and problem-posing inquiry; and functional literacy skills (including creating and reading QR codes, writing challenging, engaging clues, and properly logging and reflecting on found items).

In the perform stage, students use the knowledge and information acquired through their informational inquiries to produce or perform new work that is tied to a larger critical, conceptual, or academic goal. Within the game, students develop scavenger hunt clues for their classmates, hide them in and around their school space, and—later—search for each other's clues.

In the transform stage, students extend their performance toward publicly shared knowledge and action and focus on transforming their world. Students adapted the closed, class-only scavenger hunt into a publicly curated exhibit to affect the public's reading and interpretation of the South Central community.

Ultimately, inform, perform, and transform expands assessment to include students as both those assessed and those assessing. In allowing students to develop their own body of knowledge, with the teacher acting as facilitator in this process and encouraging transformative performative action, this process is a way to make explicit to teachers the role of student experience within classroom instruction and assessment.

This model encourages continual exploration and ensures the experiences are tied to the community and classroom needs. The assessment embeds responsiveness to the world in which the game is played and focuses understanding on the context of learning.

Antero: So this seems like a very different kind of experience for teacher and student alike.

Anansi: Absolutely. One thing I should point out is that, just as teachers through this professional development shift their roles from distributors of knowledge to facilitators of student-constructed knowledge, students, too, shift identities. In particular, I recommend allowing students to take on various roles to help them ease into the process of inquiry that Ask Anansi creates. As you'll see in this dialogue's appendix, students choose various roles for each activity. The "Checkered Flag" for instance, ensures that students complete an assignment, and the class "Diplomat" helps maintain positive working conditions between peers. These names may seem silly or "weird" for students, but assigned role shifts (even temporarily) help move students toward tangible research results and build ownership on specific components of the work within the classroom.

Antero: I think I'm still not clear how all of this stuff happening in individual classrooms has anything to do with teacher professional development and building teacher communities.

Anansi: I guess that the web I've been weaving still has a few gaps in it, doesn't it? Imagine for a second that the professional development that teachers have been encountering for eons (at least in spider years) no longer exists. Instead, teachers walk into a space that is collaboratively productive, and they take turns posing questions and engaging in dialogue with each other. For a multidisciplinary space, one can imagine that the questions will mimic student questions. The science teacher may not understand the principles that the art teacher is hoping to teach. What happens in this space is that teachers co-construct a series of question-based objectives for their individual classrooms. They do this through engagement and provocation from their peers. In this way, each teacher develops a model that meets the nuanced contexts of their classroom communities and builds a stronger relational component to their professional development experience. By yielding ownership over the space to teachers and participatory experiences, school administrators ensure that more attention is placed on student needs and a stronger network of knowledge production is formed among the teaching staff.

Antero: Okay, I think I'm getting this. But can you clarify how you ensure that Ask Anansi remains both relevant and flexible?

Anansi: Although Ask Anansi was piloted in a ninth-grade classroom, its premise extends beyond a single age group. In its most basic sense, Ask Anansi invites students to ask and explore questions of their own design. There is no reason Ask Anansi cannot be developed for students of all grades. The principles of storytelling and personal inquiry translate across ages.

As explained above, the participatory professional development that leads teachers to use Ask Anansi within a classroom starts with a basic question that is distilled into the specific needs of a classroom. By bringing student inquiry to bear on a topic through Ask Anansi, the classroom community of students will ensure relevancy.

Antero: Okay, I'm willing to try this with a group of teachers at my school, but how will we be able to sustain this project?

Anansi: The long-term sustainability of Ask Anansi relies on teacher and administrator collaboration. This game does not require a specific textbook, daily practice exam drills, or other components of a standardized-testing climate. Instead, authentic learning experiences are drawn from the community around students in ways that provoke standards-supporting English language arts instruction. This moves teacher professional development beyond the climate of high-stakes testing.

Antero: It would really help me if you could show me what the Ask Anansi goals look like in a hypothetical setting.

Anansi: Here goes: Ask Anansi's goal is to guide students toward collective inquiry around a negotiated topic and civic engagement by addressing the underlying causes of these topics. For example, a class may investigate why the food at their school is unpopular. By researching nutrition, budgeting, and distribution of food and doing qualitative surveys and an ethnographic analysis of student perceptions of school food, students may determine that a lack of variety (due to budget and contracting constraints) and a social perception (that the food is "bad") are detracting from students receiving adequate nutrition during the day. Next, students may determine that a course of action is to develop a coalition of concerned parents and students, speak at school board meetings, and even stage a cafeteria sit-in. Students will reflect on their efforts, discuss changes they have made, and record these steps in text messages, video, and mapping applications on mobile devices.

Although the main products of this game are problem posing, critical thinking, and civic participation, the goal of the game is based in the alternate reality game's fiction: they must satisfy the insatiable need of Anansi for a good story. Asking Anansi a question seems innocuous. A comment is fed to the class's spider to communicate to Anansi: "Why is the food at our school so bad?" A day later, a response is given cryptically—coordinates to the loading dock for the school's food shipments and a riddle guiding students to photograph and blog about the nutritional information they can ascertain.

The game's initial premise of asking a simple question has significant repercussions. Anansi will not simply provide an answer. He tricks, confounds, and teases students. Anansi's messages are often shrouded as riddles, QR codes, and even latitude and longitude coordinates that need to be determined and visited. Like the media messages that students are challenged to assess critically, Anansi's dialogue with students challenges concerns of power, dominance, and agency in a capitalist environment.

As students gain more information, Anansi's responses become more demanding. Students will regularly dialogue and blog about their experiences. Anansi may hack or edit their information in an effort to further a good story.

After students have completed their initial research and analysis, Anansi tells students that they have the pieces of a great story but need to weave them into action. Students need to begin working toward a course of action around the information they have received. Collective action and models of engagement are examined by the class, and a strategic plan is developed and enacted.

Anansi confesses at the end of the game that he tricked students in places with his difficult clues. He suggests that the students should recruit others to continue the story they have weaved together. After all, Anansi reminds players, a story never really ends. We may continue to talk about what happens until the next series of adventures. An inquiry and effort to improve food quality at a school is an ongoing endeavor that students will continue to work on.

Antero: You are one smart spider!

Anansi: Thank you.

Antero: I was wondering something. Even though you are saying most of this work will be constructed by teachers within their professional development, do you have some worksheet models to help us get the ball rolling?

Anansi: I suppose I could allow you to take a look. On the count of three, I will disappear, and a worksheet will take my place. (I assure you this is a painless process for me.) One, two, THREE!

The following two worksheets illustrate specific examples of assessment production tools within an alternate reality game classroom experience. Although readers are welcome to duplicate these tools, they are reproduced here primarily as examples of ways to utilize alternate reality game experiences within academic settings.

Ask Anansi Brainstorming Activity: Asking a Question

Name: ______________________________________ Date: __________

1. Over the past week, you and your classmates have investigated the background and identity of Anansi. Please list three characteristics of Anansi that you have discovered, and describe how you know this information.

a. Characteristic: ____________________ Source: __________________

b. Characteristic: ____________________ Source: __________________

c. Characteristic: ____________________ Source: __________________

2. Today, as a class you are going to come up with a question to ask Anansi to answer for you. This question should relate to this community and your day-to-day experiences here at SCHS.

a. Please select classmates to perform the following roles:

Diplomat (helps facilitate work between group members)

Engineer (oversees construction of work)

Checkered Flag (keeps group working in a timely manner)

Tourist (looks at the work from the view of the "other")

Portal (sends the question to Anansi at 323-xxx-xxxx)

b. Discuss with your classmates the two issues you have been thinking about, and as a group, choose one question. Write the question here:

____

____

____

c. After you are done, write your reflections on this process on the back of this paper. What did you think about how your group interacted? What was challenging? Why did the class think the question you chose was important? What do you think Anansi will say?

Ask Anansi Communication Log

Name ______________________________

Date and time message was sent: ______________________________

Portal: ______________________________

Sent message: ______________________________

____

____

____

What did I think Anansi would say?

____

____

____

Was a response received? Yes No

Date and time when a message was received:

What did Anansi actually say?

____

____

____

What do I think this means?

____

____

What questions do I have? What words or ideas are unclear?

____

____

____

Class next steps:

____

____

____

Appendix D: A Framework for Wireless Critical Pedagogy

Although this book presents research into using alternate reality games in an urban high school for a general audience, I engaged in this investigation to address the learning, professional development, and career-long needs of classroom teachers. Some aspects of this work have appeared in academic journals elsewhere. Below, I offer a summary of a framework for teacher development and learning based on the research presented in this book. This framework for wireless critical pedagogy places students' individualized learning needs first. Technology aids in places, but this framework—built off the lessons learned in running Ask Anansi in my classroom during this study and in my years of teaching experience—does not require expensive devices or lightning speed and unfiltered Internet. Instead, it requires mutual respect, community-driven pedagogical goals, and a willingness for teachers to adjust their practice based on the students in their classrooms and their sociocultural lived experiences. On the one hand, this framework outlines ways that mobile devices may be harnessed in classrooms not just academically but for the purpose of developing critical consciousness. On the other hand, this framework also sustains in-class connected learning without these devices.

This wireless critical pedagogy looks to build opportunities for student empowerment, access to career-preparing skill sets, and distribution of youth counternarratives within the school community. Within this vision of a modern critical pedagogy, mobile devices can offer a pragmatic means for differentiation and socialization, addressing what Thomas Philip and I have called the "3Ts" of technology.[1] By helping to develop empowered identities via the mobile opportunities of differentiation and socialization, these devices can support what Paulo Freire calls "conscientization"—a critical consciousness of the social and political elements of students' world and the ability to act on these elements.[2] Below I detail both of these opportunities and then discuss them within the larger framework for pedagogical support in classrooms with regard to technology.

Differentiation through mobile devices means personalizing learning experiences for students in ways that respond to the needs of a particular classroom. For instance, some of the ways I differentiated learning in my own classroom that should be considered by other educators include the following:

- **Audio engagement with texts:** This shifts schooling away from privileging only print media.
- **Varied opportunities for textual production:** Producing images, video reflections, interviews, and text messages varies what "work" looked like within my classroom.
- **Graphic organizers and supports:** Mindmaps, for instance, support students in thinking systemically and making connections across personal experiences and academic curriculum.
- **Continual, formative assessment of student learning:** By asking students to respond to text messages while in class, for instance, I was able to gauge individual student learning and engagement and revise lesson plans and instruction based on responses students sent to me or feedback they provided.
- **Expanded resources and methods for independent checking for understanding:** Through online searches and peer collaboration, the number of resources for furthering the scope of student research and investigation quickly increased. Likewise, students' reflections on their own learning were multimodal and varied. They could mindmap, record, photograph, write, and text their experiences, highlights, and challenges within the classroom, taking greater ownership and responsibility for the onus of critical research within their classroom.

Recognizing the many ways that student learning experiences were differentiated, the iPods in my own classroom helped to realign the text, tools, and talk that were supported within my classroom. Further, mobile devices in our classroom helped enrich the in-class community and extended this community beyond the typical hours of the school day. In doing so, these devices extended learning beyond the realm of the classroom walls in ways that traditional homework has not. Mirroring the connected learning forms of peer-supported and interest-driven learning, below are several ways that mobile media devices acted as tools for socialization:

- **Enable communication between mobile users:** Peer-to-peer communication meant collaboration on work for class took place at home, over the weekend, and sometimes during lunch periods.
- **Signal student dispositions and identity practices:** By personalizing their devices (as described in chapter 3), students were able to display and acknowledge aspects of their identity that were embraced by their peers.
- **Foster learning opportunities beyond the walls of the classroom:** Students interviewed community members through text and chat applications, called local businesses for opinions, and documented their community through photography and documentary production.
- **Bridge in-class experiences with civic opportunities for learning:** Students had increased service learning opportunities both in and out of school.[3]
- **Develop technical literacies:** Through increasing student confidence with mobile media apps and digital textual production, students saw themselves as a community of researchers, utilizing technology to increase advocacy.

Looking at the above examples of socialization and differentiation, it is worth reflecting on how these activities mirror connected learning and the cultural shifts we see today with regards to student engagement with technology. The components of wireless critical pedagogy listed below highlight the ways that classrooms must become more student centered. Mobile devices can emphasize teachers' responses to students. Creating content, posing questions, and building theory, students share their work with each other and their teacher, all of whom can provide feedback, suggestions, and critiques. Additionally, the possibilities of community and civic engagement that digital tools offer signal how a wireless critical pedagogy may leverage participatory media tools to reach *across* the boundaries between school and community to comingle the locations of the student learning experience both physically and virtually. That is, students can physically interact and document learning in schools and communities, and they also can connect to individuals through mobile tools for virtual communication, interaction, and provocation.

The tools for both independence and community-driven research and engagement here act as the fulcrum for a critical pedagogy in the shifting culture of participatory, networked media. This critical pedagogy leverages wireless tools to build humanizing relationships within the classroom and also to extend the reaches of this classroom into the surrounding community. Looking at the many ways that mobile media can amplify student agency, voice, and community within traditional classrooms, these devices can be utilized in ways that challenge dominant power structures in schools. However, in doing so, a negotiation between teachers and students must be reached. Classroom community and students needs come first, and technology second.

Below, I chart the key tenets of a wireless critical pedagogy. This chart takes into account the kinds of differentiation, socialization, and instructional shifts that can occur in classrooms today, regardless of the digital tools present. This is a *wireless* critical pedagogy both because it can emphasize the cultural shifts occurring as a result of new technologies and also because it means that powerful critical instruction is not tethered to devices that need to be plugged in or charged when running low on battery life. This list is not exhaustive. It is a starting place for teacher conversations and reflections on their own practices.

Table D.1

Key components of wireless critical pedagogy

Student centered	• Student interest, knowledge, and perspective drive content and production. • Teachers and students collaborate to engage in youth participatory action research.
Empowered identities	• Mobile devices can allow students to document, share, and amplify their expertise and thereby act as tools for Freire's concept of "conscientization." • Likewise, adjusting the activities within classrooms to situate student learning within various roles shifts how students perceive and interpret class work and its relevance in the "real world."
Community driven and responsive	• In conjunction with youth-driven research practices, work within classroom contexts speaks to and focuses on community needs and concerns. • Critical educators can help bridge in-class learning with the expertise, opportunities, and challenges that are faced beyond the school boundaries. Through digital tools, visits, and role-play, alternate voices help bolster student interactions with their communities.
Culturally relevant	• Although mobile devices can be seen as ubiquitously embraced as part of youth cultural practices, bringing these devices into classrooms changes the context of how they are perceived. Simply "using phones" in a classroom is not culturally relevant. • Applying youth cultural practices—including student personalization of mobile devices and fluidity of social and academic time within classrooms—responds to and builds on the ways mobile devices are interpreted in youth social interactions.
Critical technical and academic literacies	• Classroom learning still focuses on academic literacies and technical skills. However, these are applied within purposeful contexts. Students produce academic texts and develop technically complex media in order to advocate, inform, persuade, and ignite discussion among an audience. • Although students still write and produce research reports, persuasive essays, and other content expected within traditional classrooms, their work can look different. A student's essay may be a persuasive memo written to (and actually given to) the city council. A research report may be turned into an edited segment of a *Wikipedia* entry. A response to a literary text may become a blogpost to engage in public-driven discourse.
Not reliant on technology	• Although this pedagogy is responsive to cultural shifts as a result of participatory media tools like mobile devices, it does not require expensive technology. • A wireless critical pedagogy revitalizes critical pedagogy for the twenty-first century. It does not simply use digital tools within a classroom. Educators need to look beyond specific tools and apps to focus on incorporating the cultural practices of participatory culture for critical education.

Notes

Introduction

1. National Commission on Excellence in Education, *A Nation at Risk: The Imperative for Educational Reform* (Washington, DC: U.S. Government Printing Office, 1983).

2. D. Guggenheim, director, *Waiting for Superman*, film, Paramount Vantage, 2010.

3. A. Lenhart, "Teens and Mobile Phones over the Past Five Years: Pew Internet Looks Back," Pew Research Center, August 19, 2009; A. Lenhart, K. Purcell, A. Smith, and Z. Zickuhr, "Social Media and Mobile Internet Use among Teens and Young Adults," Pew Research Center, 2010.

4. T. L. Friedman, *The World Is Flat: A Brief History of the Twenty-first Century* (New York: Picador, 2005); C. Shirky, *Here Comes Everybody: The Power of Organizing without Organizations* (New York: Penguin Press, 2008).

5. U.S. Department of Education, *A Blueprint for Reform: The Reauthorization of the Elementary and Secondary Education Act* (Alexandria, VA: Education Publications Center, U.S. Department of Education, 2010).

6. J. Duncan-Andrade, "Note to Educators: Hope Required When Growing Roses in Concrete," *Harvard Educational Review* 79(2) (2009): 181–194.

7. A detailed discussion of my research methodology is found in appendix A of this book.

8. H. Jenkins, *Convergence Culture: Where Old and New Media Collide* (New York: New York University Press, 2008); H. Jenkins, K. Clinton, R. Purushotma, A. J. Robison, and M. Weigel, *Confronting the Challenges of Participatory Culture: Media Education for the Twenty-first Century* (Chicago: MacArthur Foundation, 2009); H. Jenkins, S. Ford, and J. Green, *Spreadable Media: Creating Value and Meaning in a Networked Culture* (New York: New York University Press, 2013).

9. Jenkins et al., *Confronting the Challenges*.

10. C. Schuler, *Pockets of Potential: Using Mobile Technologies to Promote Children's Learning* (New York: Joan Ganz Cooney Center, Sesame Workshop, 2009).

11. D. De Vise, "Wide Web of Diversions Gets Laptops Evicted from Lecture Halls," *Washington Post*, March 9, 2010, A1; N. Frey and D. Fisher, "Doing the Right Thing with Technology," *English Journal* 97(6) (2008): 38–42; International Center for Media and the Public Agenda, *A Day without Media* (College Park: University of Maryland, 2010).

12. Lenhart et al., *Social Media and Mobile Internet Use*.

13. A. Perrin and M. Duggan, "Americans' Internet Access: 2000–2015," Pew Research Center, 2015.

14. National Telecommunications and Information Administration, "Falling through the Net II: New Data on the Digital Divide," NTIA, Washington, DC, 1998; National Telecommunications and Information Administration, "Falling through the Net: Defining the Digital Divide," NTIA, Washington, DC, 1999; National Telecommunications and Information Administration, "Falling through the Net: Toward Digital Inclusion: A Report on Americans' Access to Technology Tools," NTIA, Washington, DC, 2000.

15. J. Light, "Rethinking the Digital Divide," *Harvard Educational Review* 71(4) (2001): 709–733.

16. Ibid., 712.

17. L. Cuban, *Teachers and Machines: The Classroom Use of Technology since 1920* (New York: Teachers College Press, 1986).

18. Ibid., 109.

19. Cuban, *Teachers and Machines*.

20. M. Ito et al., *Connected Learning: An Agenda for Research and Design* (Irvine, CA: Digital Media and Learning Research Hub Reports on Connected Learning, 2013).

21. I have highlighted examples of classroom-based connected learning in A. Garcia, ed., *Teaching in the Connected Learning Classroom* (Irvine, CA: Digital Media and Learning Research Hub, 2014).

22. Ito et al., *Connected Learning*, 7.

23. Garcia, *Teaching in the Connected Learning Classroom*.

24. R. C. Pfister, *Hats for House Elves: Connected Learning and Civic Engagement in Hogwarts at Ravelry* (Irvine, CA: Digital Media and Learning Research Hub, 2014).

25. Silicon Valley entrepreneur and cofounder of PayPal Peter Thiel annually invests in twenty young people under age twenty-three to avoid or drop out of college and

focus on their own passions, business ideas, and learning pathways. Thiel Foundation, The Thiel Fellowship, 2015, http://thielfellowship.org.

26. Jenkins, *Convergence Culture.*

27. C. Bonk, *The World Is Open: How Web Technology Is Revolutionizing Education* (San Francisco: Jossey-Bass (2009); B. Plester, C. Wood, and B. V. Bell, "Txt Msg N School Literacy: Does Texting and Knowledge of Text Abbreviations Adversely Affect Children's Literacy Attainment?," *Literacy* 42(3) (2008): 137–144; Schuler, *Pockets of Potential.*

28. Cuban, *Teachers and Machines.*

29. T. M. Philip and A. Garcia, "Schooling Mobile Phones: Assumptions about Proximal Benefits, the Challenges of Shifting Meanings, and the Politics of Teaching," *Educational Policy* 29(4) (2014): 676–707; T. M. Philip and A. Garcia, "The Importance of Still Teaching the iGeneration: New Technologies and the Centrality of Pedagogy," *Harvard Educational Review* 83(2) (2013): 300–319.

30. M. Ito et al., *Hanging Out, Messing Around, and Geeking Out: Kids Living and Learning with New Media* (Cambridge, MA: MIT Press, 2009).

31. J. Chafel, "Schooling, the Hidden Curriculum, and Children's Conceptions of Poverty," *Social Policy Report* 11(1) (1997): 1–18.

32. D. Massey, *Categorically Unequal: The American Stratification System* (New York: Russell Sage Foundation, 2007).

33. C. L. Coghlan and D. W. Huggins, "That's Not Fair!" A Simulation Exercise in Social Stratification and Structural Inequality," *Teaching Sociology* 32(2) (2004): 177–187; S. Kleinman and M. Copp, "Denying Social Harm: Students' Resistance to Lessons about Inequality," *Teaching Sociology* 37 (2009): 283–293; D. Leclerc, E. Ford, and E. J. Ford, "A Constructivist Learning Approach to Income Inequality, Poverty and the 'American Dream,'" *Forum for Social Economics* 38(2–3) (2009): 201–208.

34. A. Lenhart, "Teens, Smartphones and Texting," Pew Research Center, 2012.

35. J. P. Gee, *New Digital Media and Learning as an Emerging Area and "Worked Examples" as One Way Forward* (Cambridge, MA: MIT Press, 2010); Jenkins et al., *Confronting the Challenges*; S. Livingstone, *Children and the Internet* (Cambridge: Polity, 2009).

36. Garcia, *Teaching in the Connected Learning Classroom*; A. Garcia, "Trust and Mobile Media Use in Schools," *Educational Forum* 76(4) (2012): 430–433; A. Garcia, "Inform, Perform, Transform: Modeling In-School Youth Participatory Action Research through Gameplay," *Knowledge Quest* 41(1) (2012): 46–52; A. Garcia, "Like Reading" and Literacy Challenges in a Digital Age," *English Journal* 101(6) (2012): 93–96.

37. M. Gold and G. Braxton, "Considering South-Central by Another Name," *Los Angeles Times*, April 10, 2003, B3.

38. P. Lipman, *High Stakes Education: Inequality, Globalization, and Urban School Reform* (New York: Routledge, 2003).

Chapter 1

1. D. Solorzano, M. Ceja, and T. Yosso, "Critical Race Theory, Racial Micro Aggressions and Campus Racial Climate: The Experiences of African American College Students," *Journal of Negro Education* 69 (2000): 60–73.

2. Ibid., 60.

3. California Department of Education, *California Basic Educational Data System*, 2010.

4. J. Rogers, S. Fanelli, R. Freelon, D. Medina, M. Bertrand, and M. Del Raso, *Educational Opportunities in Hard Times: The Impact of the Economic Crisis on Public Schools and Working Families*, a California Educational Opportunity Report (Los Angeles: UCLA IDEA, UC/ACCORD, 2010).

5. D. Guggenheim, director, *Waiting for Superman*, film, Paramount Vantage, 2010.

6. Rogers et al., *Educational Opportunities*.

7. T. Coates, *Between the World and Me* (New York: Spiegel and Grau, 2015).

8. M. Alexander, *The New Jim Crow: Mass Incarceration in the Age of Colorblindness* (New York: New Press, 2012).

9. M. Desmond, *Evicted: Poverty and Profit in the American City* (New York: Crown, 2016).

10. My decision to ignore the policy requiring students who arrived after the bell rang to wait in the tardy line was driven by a belief that students' time is best used within class. Often the same students were tardy day in and day out, and for them, detention notices and calls home to a disconnected number did not function as the deterrent they are designed to be. Later in this chapter, I look at how inconsistencies in tardy policy shaped my decision to welcome students into my class.

11. See also J. Duncan-Andrade, "Note to Educators: Hope Required When Growing Roses in Concrete," *Harvard Educational Review* 79(2) (2009): 181–194.

12. A. Garcia, "Containing Olive: Restraining a Dog's Wild Heart and the Plight of Student Nature," *The American Crawl*, 2012, http://www.theamericancrawl.com/?p=1090.

13. With "standard repertoires" here, I'm alluding to Gutiérrez and Rogoff's framing of "repertoires of linguistic practices" in K. D. Gutiérrez and B. Rogoff, "Cultural Ways of Learning: Individual Traits and Repertoires of Practice," *Educational Researcher* 32(5) (2003): 19–25. See also M. Orellana, C. Lee, and D. Martínez, "More Than Just a Hammer: Building Linguistic Toolkits," *Issues in Applied Linguistics* 18(2) (2010): 181–187.

14. Although not the focus of my initial study, the adults wielding the language of "ladies and gentlemen" were primarily white and primarily male. Alongside the language of control is a racial and gendered reading of power in urban schools today.

15. M. Foucault, *Discipline and Punish: The Birth of the Prison* (1975) (New York: Vintage, 1995), 266.

16. A. S. Bryk, L. M. Gomez, A. Grunow, and P. G. LeMahieu, *Learning to Improve: How America's Schools Can Get Better at Getting Better* (Cambridge, MA: Harvard Education Press, 2015).

17. California Department of Education, "The *Williams* Case: An Explanation," 2015, http://www.cde.ca.gov/eo/ce/wc/wmslawsuit.asp.

18. National Governors Association Center for Best Practices, Council of Chief State School Officers, *Common Core State Standards* (Washington, DC: National Governors Association Center for Best Practices, Council of Chief State School Officers, 2010).

19. Los Angeles Unified School District, "School Report Card," name of high school removed, 2011–2012.

20. M. Ito, *Personal, Portable, Pedestrian: Mobile Phones in Japanese Life* (Cambridge, MA: MIT Press, 2006); C. Schuler, *Pockets of Potential: Using Mobile Technologies to Promote Children's Learning* (New York: Joan Ganz Cooney Center, Sesame Workshop, 2009).

21. S. A Phillips, *Wallbangin': Graffiti and Gangs in L.A.* (Chicago: University of Chicago Press, 1999); M. Sanchez-Tranquilino, "Space, Power, and Youth Culture: Mexican American Graffiti and Chicano Murals in East Los Angeles, 1972–1978), in *Looking High and Low: Art and Cultural Identity*, edited by B. J. Bright and L. Bakewell, 55–88 (Tucson: University of Arizona Press, 1995).

21. See data on teacher retention, including National Commission on Teaching and America's Future, "The High Cost of Teacher Turnover," Policy brief, NCTAF, Washington, DC, 2007.

Chapter 2

1. D. Rushkoff, *Present Shock: When Everything Happens Now* (New York: Current, 2013).

2. A. Smith, "U.S. Smartphone Use in 2015," Pew Research Center, 2015, http://www.pewinternet.org/2015/04/01/us-smartphone-use-in-2015/.

3. Ibid., 39.

4. A. Lenhart, "Teens, Smartphones and Texting," Pew Research Center, 2012.

5. A. Perrin and M. Duggan, "Americans' Internet Access: 2000–2015," Pew Research Center, 2015.

6. H. Jenkins, K. Clinton, R. Purushotma, A. J. Robison, and M. Weigel, *Confronting the Challenges of Participatory Culture: Media Education for the Twenty-first Century* (Chicago: MacArthur Foundation, 2009).

7. A "nutrition period" is a fifteen-minute mid-morning break that gives students an opportunity to get breakfast through the district's free and reduced-price meal plan.

8. S. Turkle, *Reclaiming Conversation: The Power of Talk in a Digital Age* (New York: Penguin Press, 2015); S. Turkle, *Alone Together: Why We Expect More from Technology and Less from Each Other* (New York: Basic Books, 2011).

9. See D. Rose, *Enchanted Objects: Innovation, Design, and the Future of Technology* (New York: Scribner, 2014).

10. For extended analysis of the role of sexting, see d. boyd, "Teen Sexting and Its Impact on the Tech Industry," paper presented at the Read Write Web 2WAY Conference, New York, NY, June 13, 2011, http://www.zephoria.org/thoughts/archives/2011/06/16/teen-sexting-and-its-impact-on-the-tech-industry.html; d. boyd, *It's Complicated: The Social Lives of Networked Teens* (New Haven, CT: Yale University Press, 2014); D. Sacco, R. Argudin, J. Maguire, and K. Tallon, "Sexting: Youth Practices and Legal Implications," Berkman Center Research Publication No. 2010-8, 2010.

11. C. James, *Disconnected: Youth, New Media, and the Ethics Gap* (Cambridge, MA: MIT Press, 2014).

12. A. Garcia, *Teaching in the Connected Learning Classroom* (Irvine, CA: Digital Media and Learning Research Hub, 2014).

13. d. boyd, "MySpace vs. Facebook: A Digital Enactment of Class-based Social Categories among American Teenagers," paper presented at the International Communication Association (ICA) Conference, Chicago, IL, May 23, 2009.

14. For an in-depth look at public privacy on Facebook, see d. boyd, "Networked Privacy," *Personal Democracy Forum*, video, New York, NY, June 6, 2011; boyd, *It's Complicated*.

15. N. Carr, *The Glass Cage, How Our Computers Are Changing Us* (New York: Norton, 2014); N. Carr, *The Shallows: What the Internet Is Doing to Our Brains* (New York: Norton, 2010).

16. E. B. Moje and N. Tysvaer, *Adolescent Literacy Development in Out-of-School Time: A Practitioner's Guide* (New York: Carnegie Corporation, 2010).

17. Los Angeles Unified School District, "School Report Card," name of high school removed, 2011–2012.

18. Month-to-month plans from carriers like Boost Mobile targeted urban populations, and one smartphone that was popular in the early 2000s, the Sidekick, generally was available only for this demographic.

19. J. Anyon, "Critical Pedagogy Is Not Enough: Social Justice Education, Political Participation, and the Politicization of Students," in *The Routledge International Handbook of Critical Education*, edited by M. Apple, W. Au, and L. A. Gandon, 389–395 (New York: Routledge, 2009).

20. My experiences in higher education classrooms fared worse in this regard.

Chapter 3

1. A. Collins and R. Halverson, *Rethinking Education in the Age of Technology: The Digital Revolution and Schooling in America* (New York: Teachers College Press, 2009).

2. M. Ito, *Personal, Portable, Pedestrian: Mobile Phones in Japanese Life* (Cambridge, MA: MIT Press, 2006).

3. I wasn't the only person doing such research. On the other side of Los Angeles, I was working with a team of researchers who were supporting computer science educators in utilizing mobile devices for data collection within their classes. See T. M. Philip and A. Garcia, "Schooling Mobile Phones: Assumptions about Proximal Benefits, the Challenges of Shifting Meanings, and the Politics of Teaching," *Educational Policy* 29(4) (2014): 676–707, doi: 10.1177/0895904813518105.

4. N. Mirra, A. Garcia, and E. Morrell, *Doing Youth Participatory Action Research: Transforming Inquiry with Researchers, Educators, and Students* (New York: Routledge, 2016).

5. The "Don't lose it" rule was quickly broken.

6. I know that calling someone can be done via free apps.

7. Philip and Garcia, "Schooling Mobile Phones"; T. M. Philip and A. Garcia, "The Importance of Still Teaching the iGeneration: New Technologies and the Centrality of Pedagogy," *Harvard Educational Review* 83(2) (2013): 300–319.

8. See J. Cammarota and M. Fine, eds., *Revolutionizing Education: Youth Participatory Action Research in Motion* (New York: Routledge, 2008); Mirra, Garcia, and Morrell, *Doing Youth Participatory Action Research*.

9. A. Garcia, N. Mirra, E. Morrell, A. Martinez, and D. Scorza, "The Council of Youth Research: Critical Literacy and Civic Agency in the Digital Age," *Reading and Writing Quarterly* 31(2) (2015): 151–167.

10. J. Lave and E. Wenger, *Situated Learning: Legitimate Peripheral Participation* (Cambridge, UK: Cambridge University Press, 1991).

11. Philip and Garcia, "Schooling Mobile Phones"; Philip and Garcia, "The Importance of Still Teaching the iGeneration."

12. N. Selwyn, "Exploring the 'Digital Disconnect' between Net Savvy Students and Their Schools," *Learning, Media and Technology* 31(1) (2006): 5–17.

13. "City Schools to Lift Cell Phone Ban," *Public School Press*, January 2015, pp. 1–2, New York City Department of Education, http://schools.nyc.gov/NR/rdonlyres/D0C25137-9C49-4330-8DAD-54E83248E14C/0/PSP_January2015.pdf.

14. Associated Press, "School Phone Bans Are Booming Business for NYC's Cell-Storing Trucks," *New York Daily News*, October 4, 2012, http://www.nydailynews.com/new-york/school-phone-bans-big-business-nyc-cell-storing-trucks-article-1.1174705.

15. J. P. Gee, *Situated Language and Learning: A Critique of Traditional Schooling* (New York: Routledge, 2004).

16. A. Garcia, "Rethinking MySpace: Using Social Networking Tools to Connect with Students," *Rethinking Schools* 22(4) (2008): 27–29.

17. d. boyd, *It's Complicated: The Social Lives of Networked Teens* (New Haven, CT: Yale University Press, 2014).

18. M. C. Hagood, "A Rhizomatic Cartography of Adolescents, Popular Culture, and Constructions of Self," in *Spatializing Literacy Research and Practice*, edited by K. M. Leander and M. Sheehy, 143–160 (New York: Peter Lang, 2004); K. M. Leander and M. Sheehy, eds., *Spatializing Literacy Research and Practice* (New York, Peter Lang, 2004).

Chapter 4

1. J. P. Gee, *What Video Games Have to Teach Us about Learning and Literacy* (New York: Palgrave, 2007); J. Kahne, E. Middaugh, and C. Evans, *The Civic Potential of Video Games* (Cambridge, MA: MIT Press, 2008).

2. J. Dewey, *Democracy and Education: An Introduction to the Philosophy of Education* (New York: Macmillan, 1916).

3. C. C. Abt, *Serious Games* (New York: Viking Press, 1970); I. Bogost, *Persuasive Games: The Expressive Power of Videogames* (Cambridge, MA: MIT Press, 2007); R. Koster, *A Theory of Fun for Game Design* (Scottsdale, AZ: Paraglyph Press, 2005).

4. K. Salen and E. Zimmerman, *Rules of Play: Game Design Fundamentals* (Cambridge, MA: MIT Press, 2004), 80.

5. Ibid., 475.

6. J. Huizinga, *Homo Ludens: A Study of the Play-Element in Culture* (Abingdon, UK: Routledge, 1949; New York: Routledge, 2002), 8.

7. For detailed analyses of alternate reality games, see M. Montola, J. Stenros, and A. Waern, *Pervasive Games: Theory and Design* (New York: Morgan Kaufmann, 2009).

8. J. McGonigal, "Why 'I Love Bees': A Case Study in Collective Intelligence Gaming," in *The Ecology of Games: Connecting Youth, Games, and Learning*, edited by K. Salen, 199–228 (Cambridge, MA: MIT Press, 2008).

9. M. McCall, director, *The Institute*, film, Argot Pictures, 2013.

10. Art of the H3ist, game and marketing campaign, McKinney/Silver, 2005.

11. D. Szulborski, *This Is Not a Game: A Guide to Alternate Reality Gaming* (Macungie, PA: New-Fiction Pub, 2005).

12. J. McGonigal, "'This Is Not a Game': Immersive Aesthetics and Collective Play," paper presented at the Digital Arts and Culture Conference, DAC 2003, Melbourne, 2003.

13. McGonigal, "Why 'I Love Bees.'"

14. P. Levy, *Collective Intelligence: Mankind's Emerging World in Cyberspace* (Cambridge, MA: Perseus, 1997).

15. J. McGonigal, "Gaming Can Make a Better World," talk presented at the Technology, Entertainment, Design (TED) Conference, Long Beach, CA, 2010.

16. G. Niemeyer, A. Garcia, and R. Naima, "Black Cloud: Patterns toward da Future," in *Proceedings of the Seventeenth ACM International Conference on Multimedia*, 1073–1082 (New York: ACM, 2009).

17. In the years following the Black Cloud game, the sensor technology that was developed for the game was installed and used to collect air quality data on Google's Street View cars.

18. L. Hyde, *Trickster Makes the World: Mischief, Myth, and Art* (New York: Farrar, Strauss, and Giroux, 2012).

19. Educational research would argue that Dante's behavior is partly opposing a dominant, "acting white" identity. See A. Fordham, and J. U. Ogbu, "Black Students' School Success: Coping with the 'Burden of Acting White,'" *Urban Review* 18 (1986): 176–206.

20. I use the word *enchanted* here to signal the deliberate vocabulary of "magical" devices and to highlight the shift from phones to a larger "Internet of Things." See J. Katz, *Magic in the Air: Mobile Communications and the Transformation of Social Life* (New Brunswick, NJ: Transaction Press, 2006); D. Rose, *Enchanted Objects: Innovation, Design, and the Future of Technology* (New York: Scribner, 2014).

21. Salen and Zimmerman, *Rules of Play*.

22. P. Freire and D. Macedo, *Literacy: Reading the Word and the World* (Westport, CT: Bergin and Garvey, 1987).

23. d. boyd, *It's Complicated: The Social Lives of Networked Teens* (New Haven, CT: Yale University Press, 2014).

24. J. Rogers et al., *Educational Opportunities in Hard Times: The Impact of the Economic Crisis on Public Schools and Working Families*, a California Educational Opportunity Report (Los Angeles: UCLA IDEA, UC/ACCORD, 2010).

25. G. Coleman, *Coding Freedom: The Ethics and Aesthetics of Hacking* (Princeton, NJ: Princeton University Press, 2012); C. Kelty, *Two Bits: The Cultural Significance of Free Software* (Durham, NC: Duke University Press, 2008); L. Lievrouw, *Alternative and Activist New Media* (Malden, MA: Polity, 2011).

26. E. S. Raymond, *The Cathedral and the Bazaar: Musings on Linux and Open Source by an Accidental Revolutionary* (Sebastopol, CA: O'Reilly, 2001).

27. I've considered how teachers can adopt a hacker ethos elsewhere. See A. Garcia and C. O'Donnell-Allen, *Pose, Wobble, Flow: A Culturally Proactive Approach to Literacy Instruction* (New York: Teachers College Press, 2015).

28. J. Lave and E. Wenger, *Situated Learning: Legitimate Peripheral Participation* (Cambridge, UK: Cambridge University Press, 1991).

29. Ito et al., *Hanging Out, Messing Around, and Geeking Out: Kids Living and Learning with New Media* (Cambridge, MA: MIT Press, 2009); Kelty, *Two Bits*.

30. J. P. Gee, *Situated Language and Learning: A Critique of Traditional Schooling* (New York: Routledge, 2004).

31. A. D. Smith, *Twilight: Los Angeles, 1992* (New York: Anchor Books, 1994).

Chapter 5

1. My colleagues and I have challenged the inconsistent use of the language of community in educational research. See T. M. Philip, W. Way, A. Garcia, S.

Schuler-Brown, and O. Navarro, "When Educators Attempt to Make 'Community' a Part of Classroom Learning: The Dangers of (Mis)appropriating Students' Communities into Schools," *Teaching and Teacher Education* 34 (2013): 174–183.

2. See J. Dewey, *Experience and Education* (New York: Simon and Schuster, 1938; New York: Touchstone, 1997).

3. To see how social networking existed long before the dawn of Friendster and Myspace, see T. Standage, *Writing on the Wall: Social Media—The First Two Thousand Years* (New York: Bloomsbury, 2013); T. Standage, *The Victorian Internet: The Remarkable Story of the Telegraph and the Nineteenth Century's On-line Pioneers* (New York: Bloomsbury, 1998).

4. J. Dewey, "The School and Social Process" (1897), in *The Middle Works 1889–1924*, vol. 1, *Essays on School and Society 1899–1901*, edited by J. A. Boydston, 5–20 (Carbondale: Southern Illinois University Press, 1983).

5. J. Habermas, *The Structural Transformation of the Public Sphere: An Inquiry into a Category of Bourgeois Society* (Cambridge, MA: MIT Press, 1991), 231.

6. J. Kahne and E. Middaugh, "Democracy for Some: The Civic Opportunity Gap in High School," Circle Working Paper 59, Center for Information and Research on Civic Learning and Engagement (CIRCLE), 2008.

7. S. Banaji and D. Buckingham, *The Civic Web: Young People, the Internet, and Civic Participation* (Cambridge, MA: MIT Press, 2013).

8. Ibid., 3.

9. J. Kahne, E. Middaugh, and C. Evans, *The Civic Potential of Video Games* (Cambridge, MA: MIT Press, 2008).

10. H. Jenkins, S. Ford, and J. Green, *Spreadable Media: Creating Value and Meaning in a Networked Culture* (New York: New York University Press, 2013).

11. E. Middaugh, "Service and Activism in the Digital Age: Supporting Youth Engagement in Public Life," College Civic Engagement Research Group (CERG), University of California at Riverside and Mills College, 2012.

12. C. Ruitenberg, "Educating Political Adversaries: Chantal Mouffe and Radical Democratic Citizenship Education," *Studies in Philosophy and Education* 28(3) (2009): 269–281.

13. M. Ito, *Connected Learning: An Agenda for Research and Design* (Irvine, CA: Digital Media and Learning Research Hub Reports for Connected Learning Research Network, 2013).

14. T. M. Philip and A. Garcia, "Schooling Mobile Phones: Assumptions about Proximal Benefits, the Challenges of Shifting Meaning, and the Politics of Teaching," *Educational Policy* 29(4) (2014): 676–707.

15. P. Freire, *Pedagogy of Hope: Reliving Pedagogy of the Oppressed* (New York: Continuum, 1994).

16. P. Freire and D. Macedo, *Literacy: Reading the Word and the World* (Westport, CT: Bergin and Garvey, 1987), 134–135.

17. See J. Duncan-Andrade and E. Morrell, *The Art of Critical Pedagogy: Possibilities for Moving from Theory to Practice in Urban Schools* (New York: Peter Lang, 2008); P. Freire, *Pedagogy of the Oppressed* (New York: Herder and Herder, 1970); P. McLaren, *Life in Schools: An Introduction to Critical Pedagogy in the Foundations of Education* (Boston: Pearson, 2007); E. Morrell, *Critical Literacy and Urban Youth: Pedagogies of Access, Dissent, and Liberation* (New York: Routledge, 2008); I. Shor, *Critical Teaching and Everyday Life* (Chicago: University of Chicago Press, 1987).

18. New London Group, "A Pedagogy of Multiliteracies: Designing Social Futures," *Harvard Educational Review* 66(1) (1996): 60–92.

19. To see a sample of this log from the past few years, visit theamericancrawl.com.

20. I've grown impatient with the narration speed on many audiobooks and tend to listen to these at double speed or faster.

21. The "banking model" of education is a key area of critique in the work of Freire and critical pedagogy. See Freire, *Pedagogy of the Oppressed*.

22. J. P. Gee, *Situated Language and Learning* (New York: Routledge, 2004).

23. N. Selwyn, "Exploring the 'Digital Disconnect' between Net Savvy Students and Their Schools," *Learning, Media and Technology* 31(1) (2006): 5–17

24. A. Garcia, "Thinking about Failure: Ways to Tell New Stories about Public Education," *DMLcentral*, 2011, http://dmlcentral.net/thinking-about-failure-ways-to-tell-new-stories-about-public-education.

25. J. S. Brown, "Growing Up Digital: How the Web Changes Work, Education, and the Ways People Learn," *Change* (March–April 2000): 10–20.

26. Ibid., 19.

27. Banaji and Buckingham, *The Civic Web*, 4.

Conclusion

1. Augustus Hawkins High School, "Schools for Community Action," 2015, http://hawkinshs-lausd-ca.schoolloop.com/SCA.

2. K. Salen, R. Torres, L. Wolozin, R. Rufo-Tepper, and A. Shapiro, *Quest to Learn: Developing the School for Digital Kids* (Chicago: MacArthur Foundation, 2011).

3. H. Blume, "LAUSD Launches Its Drive to Equip Every Student with iPads," *Los Angeles Times*, August 28, 2013,http://www.latimes.com/local/la-me-lausd-ipads-20130828-story.html.

4. H. Blume, "L.A. Students Breach School iPads' Security," *Los Angeles Times*, September 24, 2013, http://articles.latimes.com/2013/sep/24/local/la-me-lausd-ipads-20130925.

5. T. M. Philip and A. Garcia, "iFiasco in LA's Schools: Why Technology Alone Is Never the Answer," *DMLcentral*, Digital Media and Learning Research Hub, 2013, http://dmlcentral.net/ifiasco-in-la-s-schools-why-technology-alone-is-never-the-answer.

6. H. Blume, "John Deasy Resigns; Ramon Cortines Named Interim Head of L.A. Schools," *Los Angeles Times*, October 16, 2014, http://www.latimes.com/local/lanow/la-me-ln-lausd-board-terminates-deasy-contract-20141016-story.html; L. Lapowsky, "What Schools Must Learn from LA's iPad Debacle," *Wired*, May 2015, http://www.wired.com/2015/05/los-angeles-edtech.

7. H. Blume, "L.A. Unified to Get $6.4 Million in Settlement over iPad Software," *Los Angeles Times*, September 25, 2015, http://www.latimes.com/local/lanow/la-me-ln-la-unified-ipad-settlement-20150925-story.html.

8. National Governors Association Center for Best Practices, Council of Chief State School Officers, *Common Core State Standards* (Washington, DC: National Governors Association Center for Best Practices, Council of Chief State School Officers, 2010).

9. U.S. Department of Education, Race to the Top, 2014, http://www2.ed.gov/programs/racetothetop/index.html.

10. Partnership for Assessment of Readiness for College and Careers, "Sample Items," 2014, http://parcc.pearson.com/sample-items.

11. b. hooks, *Teaching Community: A Pedagogy of Hope* (New York: Routledge, 2003), 172.

12. D. Rushkoff, *Present Shock: When Everything Happens Now* (New York: Current, 2013).

13. hooks, *Teaching Community*, 173.

14. A. S. Bryk, L. M. Gomez, A. Grunow, and P. G. LeMahieu, *Learning to Improve: How America's Schools Can Get Better at Getting Better* (Cambridge, MA: Harvard Education Press, 2015), 60.

15. Ibid., 61.

16. A. Garcia, *Teaching in the Connected Learning Classroom* (Irvine, CA: Digital Media and Learning Research Hub, 2014)

17. A. Lenhart, "Teens, Smartphones and Texting," Pew Research Center, 2012; A. Perrin and M. Duggan, "American's Internet Access: 2000–2015," Pew Internet Research Center, 2015.

18. A. Garcia, "Rethinking the 'New' in 'New Media,'" *DMLcentral*, 2011, http://dmlcentral.net/rethinking-the-new-in-new-media.

19. The recent trend to pass intelligence on to our digital tools has led us to "smart" machines, "genius" features, and human-named digital companions. See A. Garcia, "Make 'Em All Geniuses: Redefining Schools, Possibilities, Equity," *DMLcentral*, 2015, http://dmlcentral.net/make-em-all-geniuses-redefining-schools-possibilities-equity.

20. B. F. Skinner, "Why We Need Teaching Machines," *Harvard Educational Review* 31(4) (1961): 377–398.

21. See L. Cuban, *Teachers and Machines: The Classroom Use of Technology since 1920* (New York: Teachers College Press, 1986); N. Selwyn, "Exploring the 'Digital Disconnect' between Net Savvy Students and Their Schools," *Learning, Media and Technology* 31(1) (2006): 5–17.

22. T. M. Philip and A. Garcia, "The Importance of Still Teaching the iGeneration: New Technologies and the Centrality of Pedagogy," *Harvard Educational Review* 83(2) (2013): 300–319; T. M. Philip and A. Garcia, "Schooling Mobile Phones: Assumptions about Proximal Benefits, the Challenges of Shifting Meanings, and the Politics of Teaching," *Educational Policy* 29(4) (2014): 676–707.

23. H. Mansfield, *The Same Axe Twice: Restoration and Renewal in a Throwaway Age* (Lebanon, NH: University Press of New England, 2000).

24. Cuban, *Teachers and Machines*.

25. A. Taylor, *The People's Platform: Taking Back Power and Culture in the Digital Age* (New York: Picador, 2014).

Appendix A

1. J. Comaroff and J. Comaroff, *Ethnography and the Historical Imagination* (Boulder, CO: Westview Press, 1992), 9.

2. J. L. Kincheloe and P. McLaren, "Rethinking Critical Theory and Qualitative Research," in *The Handbook of Qualitative Research*, edited by N. K. Denzin and Y. S. Lincoln, 303–241 (Thousand Oaks, CA: Sage, 2005).

3. F.M. Connelly and D. J. Clandinin, "Telling Teaching Stories," *Teacher Education Quarterly* 21(1) (1994): 145–158.

4. K. Anderson-Levitt, "Ethnography," in *Handbook of Complementary Methods in Education Research*, edited by J. L. Green, G. Camilli, and P. B. Elmore, 279–295 (Washington, DC: American Educational Research Association, 2006).

5. R. M. Emerson, R. I. Fretz, and L. L. Shaw, *Writing Ethnographic Fieldnotes* (Chicago: University of Chicago Press, 1995).

6. K. Charmaz, *Constructing Grounded Theory: A Practical Guide through Qualitative Analysis* (London: Sage, 2006).

7. Emerson, Fretz, and Shaw, *Writing Ethnographic Fieldnotes*.

8. R. Yin, "Case Study Methods," in *Handbook of Complementary Methods in Education Research*, edited by J. L. Green, G. Camilli, and P. B. Elmore, 111–122 (Washington, DC: American Educational Research Association, 2006).

9. D. Hymes, "Models of the Interaction of Language and Social Life" (1967), in *Sociolinguistics: The Essential Readings*, edited by C. B. Britt-Paulston and G. R. Tucker (Malden, MA: Blackwell, 2003), 38.

10. D. Barton and M. Hamilton, "Literacy, Reification and the Dynamics of Social Interaction," in *Beyond Communities of Practice: Language, Power, and Social Context*, edited by D. Barton and K. Tusting, 14–35 (Cambridge, UK: Cambridge University Press, 2005).

11. Ibid.

12. R. Rogers, E. Malancharuvil-Berkes, M. Mosley, D. Hui, and G. O'Garro Joseph, "Critical Discourse Analysis in Education: A Review of the Literature," *Review of Educational Research* 75(3) (2005): 365–416.

13. Ibid., 366.

14. N. Denzin, "The Experiential Text and the Limits of Visual Understanding," *Educational Theory* 45(1) (1995): 7–18.

15. N. Mirra, A. Garcia, and E. Morrell, *Doing Youth Participatory Action Research: Transforming Inquiry with Researchers, Educators, and Students* (New York: Routledge, 2016).

Appendix B

1. J. Schell, *The Art of Game Design: A Book of Lenses* (Burlington, MA: Morgan Kaufmann, 2008): R. Koster, *A Theory of Fun for Game Design* (Scottsdale, AZ: Paraglyph Press, 2005); A. Garcia and C. Sansing, *The Educator's Game Design Toolkit: A Game Master's Guide and Player's Guide for Teacher Professional Development* (Fort Collins, CO: CSU Ventures, 2014).

2. A. Garcia, "Teacher as Dungeon Master: Connected Learning, Democratic Classrooms, and Rolling for Initiative," in *The Role-Playing Society: Essays on the Cultural Influence of RPGs*, edited by A. Byers and F. Crocco, 164–183 (New York: McFarland, 2016).

3. R. D. Laws, *Robin's Laws of Good Game Mastering* (Austin, TX: Steve Jackson Games, 2002), 4–5.

Appendix C

1. This conversation was previously published as A. Garcia, "A Conversation with Anansi: Professional Development as Alternate Reality Gaming and Youth Participatory Action Research," in *Designing with Teachers: Participatory Approaches to Professional Development in Education*, edited by E. Reilly and I. Literat, 38–49, Digital Media and Learning Hub, USC Annenberg Innovation Lab, 2012.

2. K. Salen and E. Zimmerman, *Rules of Play: Game Design Fundamentals* (Cambridge, MA: MIT Press, 2004), 475.

3. A. McIntyre, "Constructing Meaning about Violence, School, and Community: Participatory Action Research with Urban Youth," *Urban Review* 32(2) (2000): 128.

4. Ibid., 148–149.

Appendix D

1. T. M. Philip and A. Garcia, "The Importance of Still Teaching the iGeneration: New Technologies and the Centrality of Pedagogy," *Harvard Educational Review* 83(2) (2013): 300–319.

2. P. Freire, *Pedagogy of the Oppressed* (New York: Herder and Herder, 1970).

3. For more on the role played by service learning in a digital age, see E. Middaugh, "Service and Activism in the Digital Age: Supporting Youth Engagement in Public Life," Civic Engagement Research Group (CERG), University of California at Riverside and Mills College, January 1, 2012.

Bibliography

Abt, C. C. *Serious Games*. New York: Viking Press, 1970.

Alexander, M. *The New Jim Crow: Mass Incarceration in the Age of Colorblindness*. New York: New Press, 2012.

Anderson-Levitt, K. "Ethnography." *Handbook of Complementary Methods in Education Research*, edited by J. Green, G. Camilli, and P. Elmore, 279–295. Washington, DC: American Educational Research Association, 2006.

Anyon, J. "Critical Pedagogy Is Not Enough: Social Justice Education, Political Participation, and the Politicization of Students." In *The Routledge International Handbook of Critical Education*, edited by M. Apple, W. Au and L. A. Gandin, 389–395. New York: Routledge, 2009.

Art of the H3ist. Game and marketing campaign. McKinney/Silver, 2005.

Associated Press. "School Phone Bans Are Booming Business for NYC's Cell-Storing Trucks." *New York Daily News*, October 4, 2012. http://www.nydailynews.com/new-york/school-phone-bans-big-business-nyc-cell-storing-trucks-article-1.1174705.

Augustus Hawkins High School. "Schools for Community Action." 2015. http://hawkinshs-lausd-ca.schoolloop.com/SCA.

Banaji, S., and D. Buckingham. *The Civic Web: Young People, the Internet, and Civic Participation*. Cambridge, MA: MIT Press, 2013.

Barton, D., and M. Hamilton. "Literacy, Reification and the Dynamics of Social Interaction." In *Beyond Communities of Practice: Language, Power, and Social Context*, edited by D. Barton and K. Tusting, 14–35. Cambridge: Cambridge University Press, 2005.

Blume, H. "John Deasy Resigns; Ramon Cortines Named Interim Head of L.A. Schools." *Los Angeles Times*, October 16, 2014. http://www.latimes.com/local/lanow/la-me-ln-lausd-board-terminates-deasy-contract-20141016-story.html.

Blume, H. "L.A. Students Breach School iPads' Security." *Los Angeles Times*, September 25, 2013. http://articles.latimes.com/2013/sep/24/local/la-me-lausd-ipads-20130925.

Blume, H. "L.A. Unified to Get $6.4 Million in Settlement over iPad Software." *Los Angeles Times*, September 29, 2015. http://www.latimes.com/local/lanow/la-me-ln-la-unified-ipad-settlement-20150925-story.html.

Blume, H. "LAUSD Launches Its Drive to Equip Every Student with iPads." *Los Angeles Times.*, August 28, 2013. http://www.latimes.com/local/la-me-lausd-ipads-20130828-story.html.

Bogost, I. *Persuasive Games: The Expressive Power of Videogames*. Cambridge, MA: MIT Press, 2007.

Bonk, C. *The World Is Open: How Web Technology Is Revolutionizing Education*. San Francisco: Jossey-Bass, 2009.

boyd, d. *It's Complicated: The Social Lives of Networked Teens*. New Haven, CT: Yale University Press, 2014.

boyd, d. "MySpace vs. Facebook: A Digital Enactment of Class-based Social Categories amongst American Teenagers." Paper presented at the International Communication Association (ICA) Conference, Chicago, IL, May 23, 2009.

boyd, d. "Networked Privacy." *Personal Democracy Forum*. Video. New York, NY, June 6, 2011.

boyd, d. "Teen Sexting and Its Impact on the Tech Industry." Paper presented at the Read Write Web 2WAY Conference, New York, NY, June 23, 2011. http://www.zephoria.org/thoughts/archives/2011/06/16/teen-sexting-and-its-impact-on-the-tech-industry.html.

Brown, J. S. "Growing Up Digital: How the Web Changes Work, Education, and the Ways People Learn." *Change* (March–April 2000): 10–20.

Bryk, A. S., L. M. Gomez, A. Grunow, and P. G. LeMahieu. *Learning to Improve: How America's Schools Can Get Better at Getting Better*. Cambridge, MA: Harvard Education Press, 2015.

California Department of Education. *California Basic Educational Data System (CBEDS)*. 2010.

California Department of Education. "The *Williams* Case: An Explanation." 2015. http://www.cde.ca.gov/eo/ce/wc/wmslawsuit.asp.

Cammarota, J., and M. Fine, eds. *Revolutionizing Education: Youth Participatory Action Research in Motion*. New York: Routledge, 2008.

Carr, N. *The Glass Cage: How Our Computers Are Changing Us*. New York: Norton, 2014.

Carr, N. *The Shallows: What the Internet Is Doing to Our Brains*. New York: Norton, 2010.

Chafel, J. "Schooling, the Hidden Curriculum, and Children's Conceptions of Poverty." *SRCD Social Policy Report* 11(1) (1997): 1–18.

Charmaz, K. *Constructing Grounded Theory: A Practical Guide through Qualitative Analysis*. London: Sage, 2006.

"City Schools to Lift Cell Phone Ban." *Public School Press*, January 2015, pp. 1–2. New York City Department of Education. http://schools.nyc.gov/NR/rdonlyres/D0C25137-9C49-4330-8DAD-54E83248E14C/0/PSP_January2015.pdf.

Coates, T. *Between the World and Me*. New York: Spiegel and Grau, 2015.

Coghlan, C. L., and D. W. Huggins. "'That's Not Fair!' A Simulation Exercise in Social Stratification and Structural Inequality." *Teaching Sociology* 32(2) (2004): 177–187.

Coleman, G. *Coding Freedom: The Ethics and Aesthetics of Hacking*. Princeton, NJ: Princeton University Press, 2012.

Collins, A., and R. Halverson. *Rethinking Education in the Age of Technology: The Digital Revolution and Schooling in America*. New York: Teachers College Press, 2009.

Comaroff, J., and J. Comaroff. *Ethnography and the Historical Imagination*. Boulder, CO: Westview Press, 1992.

Connelly, F. M., and D. J. Clandinin. "Telling Teaching Stories." *Teacher Education Quarterly* 21(1) (1994): 145–158.

Cuban, L. *Teachers and Machines: The Classroom Use of Technology since 1920*. New York: Teachers College Press, 1986.

Denzin, N. "The Experiential Text and the Limits of Visual Understanding." *Educational Theory* 45(1) (1995): 7–18.

Desmond, M. *Evicted: Poverty and Profit in the American City*. New York: Crown, 2016.

De Vise, D. "Wide Web of Diversions Gets Laptops Evicted from Lecture Halls." *Washington Post*, March 9, 2010, A1.

Dewey, J. *Democracy and Education: An Introduction to the Philosophy of Education*. New York: Macmillan, 1916.

Dewey, J. *Experience and Education*. New York: Simon and Schuster, 1938. Reprint, New York: Touchstone, 1997.

Dewey, J. "The School and Social Process." In *The Middle Works 1889–1924, vol. 1, Essays on School and Society 1899–1901*, edited by J. A. Boydston, 5–20. Carbondale: Southern Illinois University Press, 1983.

Duncan-Andrade, J. "Note to Educators: Hope Required When Growing Roses in Concrete." *Harvard Educational Review* 79(2) 2009: 181–194.

Duncan-Andrade, J., and E. Morrell. *The Art of Critical Pedagogy: Possibilities for Moving from Theory to Practice in Urban Schools*. New York: Peter Lang, 2008.

Emerson, R. M., R. I. Fretz, and L. L. Shaw. *Writing Ethnographic Fieldnotes*. Chicago: University of Chicago Press, 1995.

Fordham, A., and J. U. Ogbu. "Black Students' School Success: Coping with the 'Burden of Acting White'." *Urban Review* 18 (1986): 176–206.

Foucault, M. *Discipline and Punish: The Birth of the Prison*. New York: Vintage, 1995.

Freire, P. *Pedagogy of Hope: Reliving Pedagogy of the Oppressed*. New York: Continuum, 1994.

Freire, P. *Pedagogy of the Oppressed*. New York: Herder and Herder, 1970.

Freire, P., and D. Macedo. *Literacy: Reading the Word and the World*. Westport, CT: Bergin and Garvey, 1987.

Frey, N., and D. Fisher. "Doing the Right Thing with Technology." *English Journal* 97(6) (2008): 38–42.

Friedman, T. L. *The World Is Flat: A Brief History of the Twenty-first Century*. New York: Picador, 2005.

Garcia, A. "Containing Olive: Restraining a Dog's Wild Heart and the Plight of Student Nature." *The American Crawl*, 2012. http://www.theamericancrawl.com/?p=1090.

Garcia, A. "A Conversation with Anansi: Professional Development as Alternate Reality Gaming and Youth Participatory Action Research." *Designing with Teachers: Participatory Approaches to Professional Development in Education*, edited by E. Reilly and I. Literat, 38–49. Digital Media and Learning Hub, USC Annenberg Innovation Lab, Fall 2012.

Garcia, A. "Inform, Perform, Transform: Modeling in-School Youth Participatory Action Research through Gameplay." *Knowledge Quest* 41(1) (2012): 46–50.

Garcia, A. "'Like Reading' and Literacy Challenges in a Digital Age." *English Journal* 6(101): (2012): 93–96.

Garcia, A. "Make 'Em All Geniuses: Redefining Schools, Possibilities, Equity." *DMLcentral*, 2015. http://dmlcentral.net/make-em-all-geniuses-redefining-schools-possibilities-equity.

Garcia, A. "Rethinking MySpace: Using Social Networking Tools to Connect with Students." *Rethinking Schools* 22(4) (2008): 27–29.

Garcia, A. "Rethinking the 'New' in 'New Media.'" *DMLcentral*, 2011. http://dmlcentral.net/rethinking-the-new-in-new-media.

Garcia, A. "Teacher as Dungeon Master: Connected Learning, Democratic Classrooms, and Rolling for Initiative." In *The Role-Playing Society: Essays on the Cultural Influence of RPGs*, edited A. Byers and F. Crocco, 164–183. New York: McFarland, 2016.

Garcia, A., ed. *Teaching in the Connected Learning Classroom*. Irvine, CA: Digital Media and Learning Research Hub, 2014.

Garcia, A. "Thinking about Failure: Ways to Tell New Stories about Public Education." *DMLcentral*, 2011. http://dmlcentral.net/thinking-about-failure-ways-to-tell-new-stories-about-public-education.

Garcia, A. "Trust and Mobile Media Use in Schools." *Educational Forum* 76 (4) (2012): 430–433.

Garcia, A., N. Mirra, E. Morrell, A. Martinez, and D. Scorza. "The Council of Youth Research: Critical Literacy and Civic Agency in the Digital Age." *Reading and Writing Quarterly* 31(2) (2015): 151–167.

Garcia, A., and C. O'Donnell-Allen. *Pose, Wobble, Flow: A Culturally Proactive Approach to Literacy Instruction*. New York: Teachers College Press, 2015.

Garcia, A., and C. Sansing. *The Educator's Game Design Toolkit: A Game Master's Guide and Player's Guide for Teacher Professional Development*. Fort Collins, CO: CSU Ventures, 2014.

Gee, J. P. *New Digital Media and Learning as an Emerging Area and "Worked Examples" as One Way Forward*. Cambridge, MA: MIT Press, 2010.

Gee, J. P. *Situated Language and Learning: A Critique of Traditional Schooling*. New York: Routledge, 2004.

Gee, J. P. *What Video Games Have to Teach Us about Learning and Literacy*. New York: Palgrave, 2007.

Goffman, A. *On the Run: Fugitive Life in an American City*. New York: Farrar, Straus and Giroux, 2014.

Gold, M., and G. Braxton. "Considering South-Central by Another Name." *Los Angeles Times*, April 10, 2003, B3.

Guggenheim, D., director. *Waiting for Superman*. Film. Paramount Vantage, 2010.

Gutiérrez, K. D., and B. Rogoff. "Cultural Ways of Learning: Individual Traits and Repertoires of Practice." *Educational Researcher* 32(5) (2003): 19–25.

Habermas, J. *The Structural Transformation of the Public Sphere: An Inquiry into a Category of Bourgeois Society*. Cambridge, MA: MIT Press, 1991.

Hagood, M. C. "A Rhizomatic Cartography of Adolescents, Popular Culture, and Constructions of Self." *Spatializing Literacy Research and Practice*, edited by K. M. Leander and M. Sheehy, 143–160. New York: Peter Lang, 2004.

hooks, b. *Teaching Community: A Pedagogy of Hope* . New York: Routledge, 2003.

Huizinga, J. *Homo Ludens: A Study of the Play-Element in Culture*. Abingdon: Routledge, 1949. Reprint, New York: Routledge, 2002.

Hyde, L. *Trickster Makes the World: Mischief, Myth, and Art*. New York: Farrar, Straus and Giroux, 2012.

Hymes, D. "Models of the Interaction of Language and Social Life." In *Sociolinguistics: The Essential Readings*, edited by C. B. Britt-Paulston and G. R. Tucker, 30–47. Malden, MA: Blackwell, 1974.

International Center for Media and the Public Agenda. *A Day without Media*. College Park: University of Maryland, 2010.

Ito, M. *Personal, Portable, Pedestrian: Mobile Phones in Japanese Life*. Cambridge, MA: MIT Press, 2006.

Ito, M., et al. *Connected Learning: An Agenda for Research and Design*. Irvine, CA: Digital Media and Learning Research Hub Reports on Connected Learning, 2013.

Ito, M., et al. *Hanging Out, Messing Around, and Geeking Out: Kids Living and Learning with New Media*. Cambridge, MA: MIT Press, 2009.

James, C. *Disconnected: Youth, New Media, and the Ethics Gap*. Cambridge, MA: MIT Press, 2014.

Jenkins, H. *Convergence Culture: Where Old and New Media Collide*. New York: New York University Press, 2008.

Jenkins, H., K. Clinton, R. Purushotma, A. J. Robison, and M. Weigel. *Confronting the Challenges of Participatory Culture: Media Education for the Twenty-first Century*. Chicago: MacArthur Foundation, 2009.

Jenkins, H., S. Ford, and J. Green. *Spreadable Media: Creating Value and Meaning in a Networked Culture*. New York: New York University Press, 2013.

Kahne, J., and E. Middaugh. "Democracy for Some: The Civic Opportunity Gap in High School." Circle Working Paper 59. Center for Information and Research on Civic Learning and Engagement (CIRCLE), 2008.

Kahne, J., E. Middaugh, and C. Evans. *The Civic Potential of Video Games*. Cambridge, MA: MIT Press, 2008.

Katz, J. *Magic in the Air: Mobile Communications and the Transformation of Social Life*. New Brunswick, NJ: Transaction Press, 2006.

Kelty, C. *Two Bits: The Cultural Significance of Free Software*. Durham, NC: Duke University Press, 2008.

Kincheloe, J. L., and P. McLaren. "Rethinking Critical Theory and Qualitative Research." In *The Handbook of Qualitative Research*, edited by N. K. Denzin and Y. S. Lincoln, 303–342. Thousand Oaks, CA: Sage, 2005.

Kleinman, S., and M. Copp. "Denying Social Harm: Students' Resistance to Lessons about Inequality." *Teaching Sociology* 37 (2009): 283–293.

Koster, R. *A Theory of Fun for Game Design*. Scottsdale, AZ: Paraglyph Press, 2005.

Lapowsky, L. "What Schools Must Learn from LA's iPad Debacle." *Wired*, May 2015. http://www.wired.com/2015/05/los-angeles-edtech.

Lave, J., and E. Wenger. *Situated Learning: Legitimate Peripheral Participation*. Cambridge, UK: Cambridge University Press, 1991.

Laws, R. D. *Robin's Laws of Good Game Mastering* . Austin, TX: Steve Jackson Games, 2002.

Leander, K. M., and M. Sheehy, eds. *Spatializing Literacy Research and Practice*. New York: Peter Lang, 2004.

Leclerc, D., and E. Ford, and E. J. Ford. "A Constructivist Learning Approach to Income Inequality, Poverty and the 'American Dream'." *Forum for Social Economics* 38(2–3) (2009): 201–208.

Lenhart, A. "Teens and Mobile Phones over the Past Five Years: Pew Internet Looks Back." Pew Research Center, 2009.

Lenhart, A. "Teens, Smartphones and Texting." Pew Research Center, 2012.

Lenhart, A., K. Purcell, A. Smith, and K. Zickuhr. "Social Media and Mobile Internet Use among Teens and Young Adults." Pew Research Center, 2010.

Levy, P. *Collective Intelligence: Mankind's Emerging World in Cyberspace*. Cambridge, MA: Perseus, 1997.

Lievrouw, L. *Alternative and Activist New Media*. Malden, MA: Polity, 2011.

Light, J. "Rethinking the Digital Divide." *Harvard Educational Review* 71(4) (2001): 709–733.

Lipman, P. *High Stakes Education: Inequality, Globalization, and Urban School Reform*. New York: Routledge, 2003.

Livingstone, S. *Children and the Internet*. Cambridge, UK: Polity, 2009.

Mansfield, H. *The Same Axe Twice: Restoration and Renewal in a Throwaway Age*. Lebanon, NH: University Press of New England, 2000.

Massey, D. *Categorically Unequal: The American Stratification System*. New York: Russell Sage Foundation, 2007.

McCall, S., director. *The Institute*. Film. Argot Pictures, 2013.

McGonigal, J. "Gaming Can Make a Better World." Talk presented at the Technology, Entertainment, Design (TED) Conference, Long Beach, CA, 2010.

McGonigal, J. "'This Is Not a Game': Immersive Aesthetics and Collective Play." Paper presented at the Digital Arts and Culture (DAC) Conference, Melbourne, Australia, 2003.

McGonigal, J. "Why *I Love Bees*: A Case Study in Collective Intelligence Gaming." In *The Ecology of Games: Connecting Youth, Games, and Learning*, edited by K. Salen, 199–228. Cambridge, MA: MIT Press, 2008.

McIntyre, A. "Constructing Meaning about Violence School, and Community: Participatory Action Research with Urban Youth." *Urban Review* 32(2) (2000): 123–154.

McLaren, P. *Life in Schools: An Introduction to Critical Pedagogy in the Foundations of Education*. Boston: Pearson, 2007.

Middaugh, E. "Service and Activism in the Digital Age: Supporting Youth Engagement in Public Life." Civic Engagement Research Group (CERG), University of California at Riverside and Mills College, January 1, 2012.

Mirra, N., A. Garcia, and E. Morrell. *Doing Youth Participatory Action Research: Transforming Inquiry with Researchers, Educators, and Students*. New York: Routledge, 2016.

Moje, E. B., and N. Tysvaer. *Adolescent Literacy Development in Out-of-School Time: A Practitioner's Guide*. New York: Carnegie Corporation of New York, 2010.

Montola, M., J. Stenros, and A. Waern. *Pervasive Games: Theory and Design*. New York: Morgan Kaufmann, 2009.

Morrell, E. *Critical Literacy and Urban Youth: Pedagogies of Access, Dissent, and Liberation*. New York: Routledge, 2008.

National Commission on Excellence in Education. *A Nation at Risk: The Imperative for Educational Reform*. Washington, DC: U.S. Government Printing Office, 1983.

National Commission on Teaching and America's Future. "The High Cost of Teacher Turnover." Policy brief, NCTAF, Washington, DC, 2007.

National Governors Association Center for Best Practices, Council of Chief State School Officers. *Common Core State Standards*. Washington, DC: National Governors Association Center for Best Practices, Council of Chief State School Officers, 2010.

National Telecommunications and Information Administration. "Falling through the Net: Defining the Digital Divide." NTIA, Washington, DC, 1999.

National Telecommunications and Information Administration. "Falling through the Net: Toward Digital Inclusion. A Report on Americans' Access to Technology Tools." NTIA, Washington, DC, 2000.

National Telecommunications and Information Administration. "Falling through the Net II: New Data on the Digital Divide." NTIA, Washington, DC, 1998.

New London Group. "A Pedagogy of Multiliteracies: Designing Social Futures." *Harvard Educational Review* 66(1) (1996): 60–92.

Niemeyer, G., A. Garcia, and R. Naima. "Black Cloud: Patterns towards da Future." In *Proceedings of the Seventeenth ACM International Conference on Multimedia*, 1073–1082. New York: ACM, 2009.

Orellana, M., C. Lee, and D. Martínez. "More Than Just a Hammer: Building Linguistic Toolkits." *Issues in Applied Linguistics* 18(2) (2010): 181–187.

Partnership for Assessment of Readiness for College and Careers. "Sample Items." 2014. http://parcc.pearson.com/sample-items.

Perrin, A., and Duggan, M. "Americans' Internet Access: 2000–2015." Pew Research Center, 2015.

Pfister, R. C. *Hats for House Elves: Connected Learning and Civic Engagement in Hogwarts at Ravelry*. Irvine, CA: Digital Media and Learning Research Hub, 2014.

Philip, T. M., and A. Garcia. "iFiasco in LA's Schools: Why Technology Alone Is Never the Answer." *DMLcentral*, 2013. http://dmlcentral.net/ifiasco-in-la-s-schools-why-technology-alone-is-never-the-answer.

Philip, T. M., and A. Garcia. "The Importance of Still Teaching the iGeneration: New Technologies and the Centrality of Pedagogy." *Harvard Educational Review* 83(2) (2013): 300–319.

Philip, T. M., and A. Garcia. "Schooling Mobile Phones: Assumptions about Proximal Benefits, the Challenges of Shifting Meanings, and the Politics of Teaching." *Educational Policy* 29(4) (2014): 676–707. doi:10.1177/0895904813518105.

Philip, T. M., W. Way, A. Garcia, S. Schuler-Brown, and O. Navarro. "When Educators Attempt to Make 'Community' a Part of Classroom Learning: The Dangers of (Mis)appropriating Students' Communities into Schools." *Teaching and Teacher Education* 34 (2013): 174–183.

Phillips, S. A. *Wallbangin': Graffiti and Gangs in L.A.* Chicago: University of Chicago Press, 1999.

Plester, B., C. Wood, and V. Bell. "Txt Msg N School Literacy: Does Texting and Knowledge of Text Abbreviations Adversely Affect Children's Literacy Attainment?" *Literacy* 42(3) (2008): 137–144.

Raymond, E. S. *The Cathedral and the Bazaar: Musings on Linux and Open Source by an Accidental Revolutionary*. Sebastopol, CA: O'Reilly Media, 2001.

Rogers, J., S. Fanelli, R. Freelon, D. Medina, M. Bertrand, and M. Del Razo. *Educational Opportunities in Hard Times: The Impact of the Economic Crisis on Public Schools and Working Families*. A California Educational Opportunity Report. Los Angeles: UCLA IDEA, UC/ACCORD, 2010.

Rogers, R., E. Malancharuvil-Berkes, M. Mosley, D. Hui, and G. O'Garro Joseph. "Critical Discourse Analysis in Education: A Review of the Literature." *Review of Educational Research* 75(3) (2005): 365–416.

Rose, D. *Enchanted Objects: Innovation, Design, and the Future of Technology*. New York: Scribner, 2014.

Ruitenberg, C. "Educating Political Adversaries: Chantal Mouffe and Radical Democratic Citizenship Education." *Studies in Philosophy and Education* 28(3) (2009): 269–281.

Rushkoff, D. *Present Shock: When Everything Happens Now*. New York: Current, 2013.

Sacco, D., R. Argudin, J. Maguire, and K. Tallon. "Sexting: Youth Practices and Legal Implications." Berkman Center Research Publication No. 2010-8, 2010.

Salen, K., R. Torres, L. Wolozin, R. Rufo-Tepper, and A. Shapiro. *Quest to Learn: Developing the School for Digital Kids*. Chicago: MacArthur Foundation, 2011.

Salen, K., and E. Zimmerman. *Rules of Play: Game Design Fundamentals*. Cambridge, MA: MIT Press, 2004.

Sanchez-Tranquilino, M. "Space, Power, and Youth Culture: Mexican American Graffiti and Chicano Murals in East Los Angeles, 1972–1978." In *Looking High and Low: Art and Cultural Identity*, edited by B. J. Bright and L. Bakewell, 55–88. Tucson: University of Arizona Press, 1995.

Schell, J. *The Art of Game Design: A Book of Lenses* . Burlington, MA: Morgan Kaufman, 2008.

Schuler, C. *Pockets of Potential: Using Mobile Technologies to Promote Children's Learning*. New York: Joan Ganz Cooney Center, Sesame Workshop, 2009.

Selwyn, N. "Exploring the 'Digital Disconnect' between Net Savvy Students and Their Schools." *Learning, Media and Technology* 31(1) (2006): 5–17.

Shirky, C. *Here Comes Everybody: The Power of Organizing without Organizations*. New York: Penguin Press, 2008.

Shor, I. *Critical Teaching and Everyday Life*. Chicago: University of Chicago Press, 1987.

Skinner, B. F. "Why We Need Teaching Machines." *Harvard Educational Review* 31(4) (1961): 377–398.

Smith, A. "U.S. Smartphone Use in 2015." Pew Research Center, 2015. http://www.pewinternet.org/2015/04/01/us-smartphone-use-in-2015.

Smith, A. D. *Twilight: Los Angeles, 1992*. New York: Anchor Books, 1994.

Solorzano, D., M. Ceja, and T. Yosso. "Critical Race Theory, Racial Micro Aggressions and Campus Racial Climate: The Experiences of African American College Students." *Journal of Negro Education* 69 (2000): 60–73.

Standage, T. *The Victorian Internet: The Remarkable Story of the Telegraph and the Nineteenth Century's On-line Pioneers*. New York: Bloomsbury, 1998.

Standage, T. *Writing on the Wall: Social Media—The First Two Thousand Years*. New York: Bloomsbury, 2013.

Sudnow, D. *Ways of the Hand*. Cambridge, MA: MIT Press, 1993.

Szulborski, D. *This Is Not a Game: A Guide to Alternate Reality Gaming*. Macungie, PA: New-Fiction Publishing, 2005.

Taylor, A. *The People's Platform: Taking Back Power and Culture in the Digital Age*. New York: Picador, 2014.

Thiel Foundation. The Thiel Fellowship. 2015. http://thielfellowship.org.

Turkle, S. *Alone Together: Why We Expect More from Technology and Less from Each Other*. New York: Basic Books, 2011.

Turkle, S. *Reclaiming Conversation: The Power of Talk in a Digital Age*. New York: Penguin Press, 2015.

U.S. Department of Education. *A Blueprint for Reform: The Reauthorization of the Elementary and Secondary Education Act*. Alexandria, VA: Education Publications Center, U.S. Department of Education, 2010.

U.S. Department of Education. Race to the Top. 2014. http://www2.ed.gov/programs/racetothetop/index.html.

Yin, R. K. "Case Study Methods." In *Handbook of Complementary Methods in Education Research*, edited by J. Green, G. Camilli, and P. Elmore, 111–122. Washington, DC: American Educational Research Association, 2006.

Index